Joe's
expert gardening guide

create your own
small
garden

Collins

Published by Collins
An imprint of HarperCollins Publishers
Westerhill Road, Bishopbriggs, Glasgow G64 2QT
www.harpercollins.co.uk
collins.reference@harpercollins.co.uk

HarperCollins Publishers
1st Floor, Watermarque Building, Ringsend Road,
Dublin 4, Ireland

© HarperCollins Publishers 2022
Text © Joe Swift
Cover image © Joe Swift / Marianne Majerus

A catalogue record for this book is available from
the British Library

ISBN 978-0-00-846108-9

10 9 8 7 6 5 4 3 2 1

Printed in Slovenia by GPS Group

MIX
Paper from
responsible sources
FSC
www.fsc.org **FSC™ C007454**

FSC™ is a non-profit international organisation established to promote the
responsible management of the world's forests. Products carrying the FSC
label are independently certified to assure consumers that they come from
forests that are managed to meet the social, economic and ecological needs
of present and future generations, and other controlled sources.

Find out more about HarperCollins and the environment at
www.harpercollins.co.uk/green

Thanks to my agents, Charlotte Robertson and
Debbie Scheisse, and everyone at HarperCollins
Publishers including Gerry Breslin, Gordon
MacGilp, Lauren Murray and Kevin Robbins.

photo credits
page 11 © NICOLE CUBBIDGE / Alamy Stock
Photo; page 12 (top) © flowerphotos /
Alamy Stock Photo; page 13 (bottom) ©
PhotoMavenStock / Shutterstock.com; page 15
© NICOLE CUBBIDGE / Alamy Stock Photo;
page 17 © John Richmond / Alamy Stock
Photo; page 25 © Botanic World / Alamy Stock
Photo; page 28 © A Garden / Alamy Stock
Photo; page 31 © A Garden / Alamy Stock
Photo; page 33 © Pollen Photos / Alamy Stock
Photo; page 35 © Arcaid Images / Alamy Stock
Photo; page 36 (top) © Roberto La Rosa /
Alamy Stock Photo; page 36 (bottom) ©
Roberto La Rosa / Alamy Stock Photo; page 41
© Guy Bell / Alamy Stock Photo; page 43 ©
Avalon.red / Alamy Stock Photo; page 52 ©
BIOSPHOTO / Alamy Stock Photo; page 56
© A Garden / Alamy Stock Photo; page 59 ©
Andreas von Einsiedel / Alamy Stock Photo;
page 60-61 © Jacek Wac / Alamy Stock Photo;
page 65 © Anna Stowe / Alamy Stock Photo;
page 78 © Jeff Gilbert / Alamy Stock Photo;
page 79 © Ian Lamond / Alamy Stock Photo;
page 85 © Zoonar/Peter Himmelhuber / Alamy
Stock Photo; page 91 © Anna Stowe Botanica /
Alamy Stock Photo; page 94 © Jason Smalley /
Alamy Stock Photo; page 97 © Kathy deWitt /
Alamy Stock Photo; page 111 (bottom) © Anna
Kazantseva / Alamy Stock Photo; page 116 ©
Zoonar/Peter Himmelhuber / Alamy; page 117
© Richard Bryant / Alamy Stock Photo; page
118 (top) © FlowerStock / Alamy Stock Photo;
page 120 © Olga Seyfutdinova / Alamy Stock
Photo; page 126 © Botanic World / Alamy
Stock Photo; page 154 © Michael Wheatley /
Alamy Stock Photo; page 167 © Plantography
/ Alamy Stock Photo; page 174 © blickwinkel/
McPHOTO/H.-R. Mueller / Alamy Stock Photo;
page 180 © Deborah Vernon / Alamy Stock
Photo; page 185 © ronstik / Alamy Stock
Photo; page 197 © John Richmond / Alamy
Stock Photo; page 198 © Yann Avril / Alamy
Stock Photo; page 203 © Zamfir Cristian Ion /
Alamy Stock Photo; page 211 © Trevor Chriss
/ Alamy Stock Photo; page 213 © Deborah
Vernon / Alamy Stock Photo

All other images © Shutterstock.com.

Joe's

expert gardening guide

create your own
small
garden

introduction

introduction

By definition, towns and cities are densely populated
areas. As a result, green space is always at a premium
and any urbanite who has access to their own private
garden probably knows just how lucky they are.

Outside space in a city comes in all shapes and sizes.
Back and front gardens are still the most traditional,
but there are also shared communal gardens (complete
or divided up), roof terraces, balconies, front steps, side
alleys, houseboats and many other interesting spaces to
grow in. They may start off being barren and uninviting
but with imagination, creativity and application it amazes
me how quickly they can be transformed into valuable
green havens; the perfect antidote to fast city living.

Urban areas are often far detached from the natural
world, yet it's in our DNA to want to grow things and
surround ourselves with plants. The considerations are
undeniably different to those in rural settings, with issues
like privacy, boundaries and shade (cast by buildings)
coming into play and many spaces don't have any soil
at all to dig into. These are viewed as limitations, but an
exciting aspect of urban gardening is finding solutions,
making over outside spaces beyond recognition, and
breathing life into them.

There are many directions to go in and no rights or wrongs, yet with each decision and path chosen a garden is developing and your personality is being stamped onto it. Elements like paving, boundaries, containers, furniture, water and sculpture help define its structure but the plants will always play the starring role with their inimitable colours, textures, movement and fragrance, sprinkling their magic dust around to draw the seasons into our cities.

So, how do you want your small garden to feel and what kind of plants do you want to grow? Maybe an urban jungle full of dramatic architectural plants? Or packed with colour? Perhaps a soothing space for a spot of green bathing after a stressful day's work that also works nicely as a venue for inviting friends over for a drink?

Whatever you want it to look like and be used for, greening your own plot is far from being a selfish act. The wider environmental benefits are significant too, helping to reduce pollution and minimise the heat island effect (a result of the hardscape in cities). The plants will increase biodiversity, and many will be an important year-round resource for wildlife.

Gardening can be a solitary or communal act, as you choose, yet it is undeniably infectious, in turn changing neighbourhoods and the way people feel about their surroundings. One person starts on their front garden or takes on a spot of 'guerrilla gardening' by planting up the tree pit in the street and before you know it everyone's at it, swapping cuttings and seeds, talking to each other, and building a community based on their love of city green.

There are no downsides to gardening. It's good for our physical and mental health and feeds our souls. Sure, we all have some failures along the way, but learning from them is all part of the process. I guarantee you'll get far more out of making a garden than you put in and I hope this book both inspires and offers practical help for anyone looking to green up their urban space.

design
considerations

These are the key things I consider when designing city and town gardens. Keep it simple, try to tick off most of these points to form a cohesive design and you won't go far wrong.

Sitting and versatility

A key decision that can dictate the entire layout of a small space is where to sit. In the sun or shade? Near the back door or at the end of the garden? I'd rather have one large, generous and versatile (place to sit/do yoga/pot up plants/peel potatoes) area than two pinched areas neither of which feel quite generous enough.

Big things in small spaces and feature placement

A common oversight is to place (and plant) many small things in small spaces, which makes them feel busy and cluttered. Fewer and larger elements (think architectural plants/large pots/sculptural pieces) provide balance and greater impact. When placing, view the garden as an art gallery, each piece holding a space around it and perhaps framed by an arch, by plants, or by the window frames from inside. Maximise any long views where possible to draw the eye, even if this means snipping some branches off a shrub or two to see a feature. Objects are often plonked on the ground but usually look better brought up closer to eye level, perhaps raised on plinths, blocks or fixed onto walls.

imagine how different this would feel with dark paving and gravel.

Tone

How light or dark a material or plant is – as opposed to its colour – often goes overlooked but can have a huge bearing on the overall atmosphere of a garden. In dark shady spaces, generally avoid tonally dark surface materials (dark slate/stone/bark chippings) and look to lighter stone or pebbles and boundary treatments. It can make a huge difference to how airy and welcoming the garden feels and how much light is bounced back into the house.

Views from inside

If you have views from inside out into the garden, make the most of them, as during the winter months they will become priceless. A percentage of evergreen plants viewed from the inside makes a huge difference. Aim to create foreground by bringing wispy or easily trimmed (not too blocky) plants close to the window.

Movement

When one walks into any garden one is ideally drawn through it both physically (perhaps heading somewhere to sit) and visually (planting or feature compositions drawing the eye towards them). This 'movement' makes all the difference to its success. No, it doesn't mean lots going on, it's more about the placement and revealing of certain elements as one moves through. I appreciate that in tiny gardens this may be difficult to achieve, you may only have one way to go and look but making the most of it is what it's all about!

Space for planting opportunities

This is my bugbear. Builders and many landscapers love to pave right up to house and boundary walls (so you can get a broom into the corners!) and space for plants is an afterthought. Plants tend to grow better and require far less maintenance when grown in the ground (as opposed to pots) and they also bring a garden to life. If you're having paving installed, consider leaving generous gaps near the windows, along boundaries and especially in corners to soften the space and envelope yourself in plants. If you've inherited a mostly paved garden, lift a slab or two and see if you can dig out, place topsoil and plant up.

Scale, proportion and lines

Sketch it out on a simple plan or take photos and draw over them on a tablet. Look for clues, try taking lines from the windows and doors out into the garden and then some horizontally across the garden, perhaps from a fence post or brick pier, to see if they help break it up, add some proportion, connect to what's already there and help organise it in some way. If you know what size paving unit (see the Surfaces chapter) you're going to use, see it as a grid for laying over the garden plan and a good starting point. You can then apply all my other suggestions above (sitting/views/space for plants/movements and other considerations like privacy and garden style) on your sketch (use coloured pencils or an app on your tablet), which will help see it better. A design should start coming together like a phoenix rising from the ashes!

city garden style

Gardens are very personal spaces, and the most important thing is that the urban gardener loves their own plot, even if others don't so much. Objectively, however, some spaces are undeniably more successful than others. The garden maker may have an affinity with plants, a passion, and a great eye (which certainly helps), but we all have it within us to create something special and individual to us. I have seen hundreds if not thousands of gardens over the years and, for me, the key to the realisation of that dream garden is always 'focus'. Focus on where you're heading and what you're trying to achieve. Ah, it may sound easy and simple, but with gardening there are unlimited ways to go and decisions to make with tens of thousands of plants in a kaleidoscope of colours available, and umpteen materials and features to choose from. It's easy to be led down the wrong (garden) path.

Focus and keep your eye on your vision. In successful gardens, everything feels as if they're working together, the concept bold and result harmonious, and, as with all good design, greater than the sum of its parts. Coming up with a 'garden style' to aim for may sound daunting but it's a hugely exciting aspect and I want to help you to make that key decision. Choose a feel and mood to embrace and give it an identity, your identity. Over time, all gardens are tweaked and refined, they are never static nor finished but having that clear goal helps organise the garden, your time and I guarantee will save you time and money in the process. Ah, now I've got your attention!

So, what will your urban plot look like and how will it feel? This is where inspiration comes in. You may find it walking through a natural landscape like a woodland, the colours and tones of a favourite painting or perhaps a visit to other people's gardens (The National Garden Scheme charity has many fabulous gardens in your area too if you are in England and Wales). Look at books, magazines and online to pare down your options, and, sure, compile one of those 'mood boards' with images of what floats your boat. Let's not worry about the practicalities at the moment, like levels, where to sit or privacy. A design process works in layers and those decisions can all be worked into your chosen garden style as it develops.

Does your vision have an overriding theme? Perhaps an all-green garden (extremely tranquil and relaxing), a romantic space with scented climbing roses and lilies, an ultra-formal parterre with clipped topiary, or an energy-filled exotic garden packed with as much contrast and colour as you can gather? It's the most exciting decision to make and all up to you!

small gardens often create microclimates to grow a fab range of exotics!

Whichever way you go, the most important thing is that you feel comfortable in it. If you're an ultra-tidy person, you'll find it impossible to sit in a wild-style space and, vice versa, if you hanker after a wild landscape you won't feel comfortable in anything too formal. When a couple makes decisions (as a designer I've turned marriage guidance counsellor many a time!), compromises will need to be made, but try not to dilute the vision too much.

Maybe you want your garden to have a full-on international flavour like a Japanese or Mediterranean garden? Many do. I am rather wary of this approach as after a few years a garden like this may not give you the versatility you need. Sure, it can have a Japanese twist with maples, gravel and water but for longevity, I'd suggest looking for more of a hybrid or fusion style that encompasses an oriental flavour rather than a full-on Japanese garden.

Okay, now you have a vision of your garden style, everything is going to have to work with it. After 'focus', we have 'ruthless', which may require taking out or changing elements that are already there but fight against your vision. From now on, every single thing that goes into it will enhance it.

Once you know where you're heading, a design will build up in layers. It may not have a name to be boxed into but for ease let's take a 'romantic city garden' as an example. The garden may already have elements that make a nod that way (a mature apple or magnolia tree? an ivy-clad fence?) or you may need to change almost everything in it to achieve your goal. You may have inherited some hard landscaping that works or can be tweaked or adapted, by taking out a slab here

and there or adding to it with some gravel areas, but
if you're making a garden from scratch, the romantic
style will dictate choices: nothing too modern or slick.
Boundaries can be softened with climbers but if you
have a characterful old brick wall then consider leaving
at least some of it on show. Select (or make) features like
benches, arches, water features, secluded mirrors, etc.
with your romantic theme in mind.

Now we have a few elements working together, we're onto the plants that breathe life into your romantic creation. 'Focus and ruthless' combine. Forget exotics like bamboos and large-leafed palms or a wild, naturalistic grass and perennial style. Look for plants that conjure up romantic images and of course plenty of scent such as roses, mock orange (philadelphus), deutzias, hydrangeas, clematis, honeysuckles, ivy, tulips, forget-me-nots, lilies, peonies, foxgloves, hardy geraniums (as ground cover), lady's mantles and thyme for cracks in paving. Then pots, containers and troughs (terracotta or stone?) packed with scented pelargoniums and trailing ivy or perhaps a piece of topiary or two for structure? Everything layering up nicely and working towards the goal. Most plants come in a wide range of colours. An all-white planting scheme needs serious discipline but if adhered to can look wonderfully romantic or perhaps a mix of soft colours avoiding any bright clashing, energising combinations. And there we have it, the essence of a great romantic garden is born; all it needs is making! The garden style process is the same for all gardens so dream, create, play, experiment, work and focus and, believe me, it really can become a reality.

surfaces

City gardens are rarely large enough for decent-sized lawns so it's likely to be the hard surfaces (such as terraces, patios, paths and seating areas) that make your garden an inviting and practical space. They dictate the look and cost so are a key design decision. The all-important movement through the garden and the way that areas link together will be dictated by the surfaces, so they should ideally be considered early on but can be updated, retro-fitted to gardens and added to over time. Being an ex-landscaper, I build things to last but, for those who rent or are on a very tight budget, there are always shortcuts. I once built a deck out of old wooden pallets for a friend that was still going strong 5 years later.

All successful urban gardens use surfacing imaginatively. I don't mean in a fancy 'look at me' way, far from it, the best are often extremely simple and subtle. Stone, gravel and brick are the most traditional surface materials, but can be used in contemporary ways to link the garden to the architecture of the house while giving it a more 'updated' look. Look for clues in the property to link the interior and exterior such as the way bricks are laid on house or boundary walls, the direction floorboards lie inside, or the unit sizes of, say, a kitchen floor that can be found in an outdoor material.

When installing surfaces, it's essential to consider the practical issues such as levels, any steps that may need to be put in and where the water will drain off these surfaces. These issues may affect the materials you choose and where you decide to put certain areas on the ground.

combining surfaces

I rarely use more than two hard surfaces in a garden as it can look too busy. Perhaps paving or decking for the seating areas and gravel or chippings for any paths? Pick up samples and place them next to each other in the garden to imagine how they work together. Think about using two sizes of the same material to simply delineate between areas or, say, putting a band or two of granite setts between two areas as a simple detail to set them apart. (For the tone of hard surfaces, see the chapter Dealing with Shade.)

DIY or professionals?

You could consider laying your own hard surfaces – which may also be a factor in which you choose as some are easier than others as a DIY project – and therefore keep the overall cost of the garden down. Gravel and aggregate surfaces are the easiest to install, decking tends to be easier than stone or brick but still requires a certain level of carpentry skill or there are flat pack deck kits available that are a little limiting but pretty foolproof. Paving slabs and brick paving are not easy to get right the first time round and natural stone paving of varying thickness is the hardest to lay as each stone requires a different depth 'bed' of sand and cement underneath. As with many trades, a lot of the work is in the preparation of the site, in this case understanding the levels and any necessary excavation and drainage implications before embarking on the paving itself. Landscaping work is very heavy with slabs, sand and cement all requiring plenty of hard labour to move, mix and lay. This may be the area of your garden build where you may want to call in a professional landscaper to help you.

Level changes

Level changes such as steps need to be dealt with and planned at the same time as the surfaces and may have a bearing on your choice of surfacing too. In large gardens or where there are a few level changes, I often use railway sleepers as a 'riser' and then fill the 'tread' with an aggregate, which is relatively cheap and can be installed as a DIY project without concrete foundations. Steps are a little more complicated as they usually need to be tied into retaining walls either side to avoid just banking the soil, which may collapse under heavy rain. Plan well and use these kinds of landscape details to your advantage; interesting steps can make a great feature to a garden. If you use brick or paving, the shape and size of the material used will dictate the shape of the step so design the steps to fit the module size of the paving rather than ending up with lots of fiddly cuts.

Beautifully integrated and softened with plants - makes me want to explore!

Drainage

It's vital to consider the finished level of the paving and drainage when installing hard surfaces, especially when laid near to the house. Paving needs to be laid with a built-in 'runoff' of at least 1:100 (usually the bubble in a spirit level touching the line rather than being in the middle) to take the water to either a drain, a soak-away or, if the surface area of the paving isn't too great, into a well-drained planting area where it will soak back into the soil. The direction of the runoff of water needs to be thought about prior to installation, and if it is to run away from the house, plan what happens if the lawn or planting area gets saturated. Maybe your garden has a natural built-in slope that you can utilise to help the water run off?

Be careful with butting paving up to a house wall too. All buildings have a damp-proof course laid into the brickwork and any paving next to it must be a minimum of 150 mm below this level. If there is existing paving, you will need to excavate first to then build back up to this final level. If the paving is to drain towards the house and there is an existing rainwater downpipe, you could consider laying a slot drain or channel grating along the house to tie into it, which you then pave up to give a neat finish. This will need a professional landscaper to install.

Maintenance

All hard surfaces will require some maintenance, but this is often due to the aspect where the material is laid rather than the material itself. In permanently shady spots, all paving and decking will get slippery with a build-up of algae and moss over time. I find the best way to clean this is to use a pressure washer once or twice a year and blast off the slime as it doesn't involve using any nasty chemicals and is far easier and more effective than scrubbing away for ages with a stiff brush. When cleaning paving with a powerful pressure, be careful not to blast out the pointing between the slabs if it's a little cracked. Paving may from time to time need re-pointing with strong cement mix. Timber decking, whether it's softwood or hardwood will turn a silver grey over time. To bring back a hardwood natural colour, clean first and then apply a decking oil, which will help to preserve it too. Any aggregate or gravel surfaces will need replenishing every now and then to keep them looking tip-top, so when you buy the material in the first place order a few extra bags for this purpose.

surface choices

Some are easier to install than others, so on top of the cost of the material think about whether you'll need someone to help you and how much that may cost.

Concrete paving: Concrete paving comes in a huge range from regular-sized smooth slabs to quite realistic imitation stone ones that come in a variety of sizes for a random look. They are even thicknesses throughout so are quite easy to lay but will require a well-consolidated sub-base, which is not easy to DIY without some experience.

Natural stone paving: Natural stone paving is tough, beautiful and is often seen as the 'ultimate' garden material. It can look great on terraces, paths, steps etc. and will also mix well with brick or natural aggregates. Be careful not to put it near concrete or imitation paving as it will ruin the effect!

Slabs are laid like concrete paving but depending on the thickness and evenness of the stone and the way it is laid, it is one of the harder surfaces to do as a DIY project. Random rectangles in natural stone should be left to the experts. Types of stone paving include Yorkstone and other sandstones, granite, basalt, limestones, etc.

Decking: Decking is a warm, quiet surface that can work well in a more contemporary garden. The quality of the wood is key to its look, longevity and price, with treated softwoods being the cheapest and hardwoods being the most expensive. The success of a deck is in how it is designed into a garden and the surrounding planting rather than decking spread over the whole area, which looks as if it's just been plonked on the ground.

Decking is a good choice on sloping sites or where changing the drainage is an issue as the rainwater flows through it. Use a landscape fabric on the ground beneath to stop weeds coming through. This is a good DIY project.

A work of art! It's all in the detailing and finishing.

Block paving: Block paving is made from concrete and is extremely hard-wearing. It can look good in small areas or as an edging to another material but be careful using it in gardens, as large areas of block paving can make a garden look like a car park! Small areas can be laid onto sand cement on a consolidated sub-base (medium DIY skill). Larger areas can be laid onto compacted sand (whacked with a whacker plate) over a sub-base. (Small areas can make a good DIY project, but larger areas will need a professional.)

Granite setts: Granite setts come in 100x100 mm or 100x200 mm. They can be new or reclaimed. They can be used as a main surface (although the reclaimed setts can leave a rather uneven surface for seating) and are idealfor detailing such as edging, fanning out or for making paved circles. Lay in a similar way to block paving.

Brick paving: Brick paving can look fabulous for patios or paths or used to break up a large expanse of other paving. New bricks tend to be cheaper than reclaimed but make sure you use a frost-proof brick.

Lay in the same way as for block paving, but if the bricks are reclaimed and uneven, it's best to leave a gap and point or grout by hand to help even out the gaps. (Small areas can make a good DIY project, but larger areas will need a professional.)

Terracotta paving: Most terracotta paving isn't frost-proof, but there are some excellent imitation terracotta tiles that are realistic and I would recommend only using these. They look best in a small courtyard-style garden, are a warm colour and will instantly add a Mediterranean flavour to an urban garden.

If only laying a small area of these, it's best to lay a solid concrete foundation first to avoid any movement. Larger areas can be laid like concrete slabs but use a wet sand and cement mix to stop them popping up. Once the area is level, they are relatively easy to lay as they are smaller and more handleable than larger paving units.

Slate paving: Natural slate comes in various hues of greys, blue-greys and green-greys, and is extremely hard-wearing and completely stain resistant. It can look great in all styles of garden, but as with most materials, don't lay it in huge expanses without breaking it up with another material or planting. Looks great when it's wet.

Don't use thin interior tiles in the garden, only use tiles with a minimum thickness of 10 mm and lay them in the same way as a terracotta tile. Point or grout with a mortar mix to roughly match the colour of the tiles rather than a light or black colour.

Poured concrete: Poured concrete can be worked into a variety of finishes from a dull grey to an exposed aggregate or there are also colour pigments that can be added to create funky surfaces for more contemporary gardens. The area will need to be shuttered with boards first and then the concrete mixed up and poured in. The level of skill depends mainly on the finish you're looking for and most high-quality finishes will require a professional to install.

Crazy paving: Crazy paving doesn't have to be a throwback to the 1970s and can look great if well thought through and executed. It's far cheaper to buy than rectangular paving and in my opinion the larger the pieces of stone, the more practical it is and better it looks. Crazy paving is far harder to lay properly than most square paving, as the joints need to be tight for it to look good. It needs to be laid on a compacted base and each piece of paving bedded into a wet mix of sand and cement and then hand pointed. Consider laying paths with stone pieces and planting in-between with ground cover plants to soften them.

Gravel and aggregates: Gravels and aggregates such as pea shingle, Cotswold gravel, slate shale and rounded beach pebbles will add plenty of texture to a garden and can be planted through to soften it and break it up. They're not so good for seating areas but are far cheaper and quicker to install than any other paving as it is a loose material and will freely drain too. Lay landscape fabric over the ground and pin down thoroughly (there's nothing worse than seeing the fabric flapping up and giving the game away!). Cut holes and plant through where necessary and then place around 50 mm of aggregate on top. A great DIY project! Consider using a gravel stabilisation system (especially on slopes) such as honeycombed plastic modules to stop the material getting kicked or moved around over time.

Pebble mosaic: If you really want to stamp your mark on a space, how about a bespoke mosaic with pebbles? It'll need careful planning, drawing out and is slow to make but it's a way of personalising some of your paved

Victorian tiles beautifully define the seating area

areas and can work particularly well in a small courtyard garden. Lay a concrete slab and individually set pebbles into your design. Try a small area out of the way first to make sure you get the right finish before embarking on the final project.

ethics

You may find a huge disparity in prices between stone that is quarried in the UK and Europe and stone that is quarried, say, in India and China, and confusion as to the ethics behind buying such material. Find out as much information as possible when buying these materials and ask your supplier to verify that they come from reputable sources that are regularly checked for their working conditions and pay.

When buying hardwood timber for decking, ensure it has an FSC (Forest Stewardship Council) certificate to show that the wood has been logged from managed renewable sources. It isn't illegal to import or buy non-certificated wood or stone in the UK, so ask questions and only use materials you're fully comfortable with.

super simple as a
boundary or division

boundaries

Boundaries not only delineate the perimeter of your plot and generally increase privacy; in design terms, they are one of the major elements in the garden along with surfaces and built-in features. Boundaries can be subtle, helping the garden to recede into the distance, but in a more contemporary space have the potential to be part of a larger strategy to draw the eye into key areas. The smaller the garden is, the more considered the boundary treatment should be as it'll be one of the first things you clock at eye level when walking into the garden.

If new boundaries are going in from scratch as part of a redesign or as a necessity (old ones falling down?), this creates a huge opportunity to do something interesting. In truth, most of us inherit boundaries such as fences or brick walls with our gardens but it doesn't mean we're stuck with what we have; there's always a solution such as customising by staining, painting, cladding or more typically covering them with climbers – a fine way to get more greenery into a space. The term 'stone cladding' may bring frightening images of 1970s suburbia flooding back but I've recently done a project with a fabulous real stone tile product (applied to a block wall) that looks superb and works out considerably cheaper than a solid stone wall.

Budget is, of course, a huge consideration. Brick and stone walls are expensive, requiring foundation work and skilled labour to install to a high standard, but the investment can be worth it, lasting decades and giving a sense of permanence to a garden. Timber fences, hurdles and trellis are far more affordable, often off-the-peg products that are installed quickly and these

Blank walls can always be broken up with a simple feature or two.

days there are a wider range available than ever before in both traditional styles as well as more contemporary products (such as sleek louvred and slatted fence panels). Hedges can make a fine boundary even in a town or city setting as they are great for wildlife, filtering winds and reducing pollution.

Here are some of my tips when considering boundaries:

- Only install tall boundaries where you really need them rather than putting them all the way around your garden by default. Place taller sections where you particularly need to create privacy (so think about where you'll be sitting) and alternatively consider lowering the heights of a boundary to open up a good view if you have one.

- Fencing and walls often accentuate the boundary lines of a garden, thereby reinforcing its shape and footprint, but this is not always the best design approach. By using the same finish in another part of the garden (as a garden division? edging a terrace? cladding a shed?) it can help break this visual problem while bringing cohesion to the overall design.

- Solid boundaries tend not to be the best way of dealing with strong winds if you live in an exposed spot. The wind will come over the top and often create damaging eddies on the garden side. Hedges will help filter the wind as will a more 'hit and miss' fence with gaps in it.

- In shady gardens, think about how you can bounce any available light around by keeping them light tonally (painting or staining a light colour?) It can make a significant difference to the overall mood of the space, and plants often respond well to the increase in ambient light levels too.

- For continuity to the garden, try to avoid too many different boundary finishes throughout. One is usually enough but perhaps two max.

- Think about which parts of the boundary you will see. There's no point in installing a top of the range boundary all the way round and then smothering it in evergreen climbers. Invest where you can see the benefits.

- Black-stained fencing can look surprisingly effective. It can set off and intensify the colour of planting in front and works particularly well with greens.

- Use materials practically and imaginatively. Details such as the bond in a brick wall or the way the courses of stone are laid can add a level of quality and design for no extra cost, while some off-the-peg products can be customised or tweaked to look bespoke.

creating privacy

Privacy is a key factor in a town or city as it's difficult to fully unwind when you feel as if you're being spied upon (even if you're not).

A common approach – and in my opinion a big mistake – is to turn the garden into a fortress by building up the trellising on top of the fence as high as possible or growing tall hedges to block everything out. This can result in a very shady garden and work out costly too as either expensive structures or plants are bought for instant results. If very fast-growing plants are chosen (such as leylandii cypress trees or 'mile a minute' Russian vine), this can lead to many issues later (shade/dry beneath/unmanageable/angry neighbours!).

The fortress approach is more often than not completely unnecessary. Consider a more strategic approach to the problem and try and integrate it into the overall design. First, work out where you want to spend time in the garden and precisely where you may want to increase privacy. This will dictate the size and height of the plant or structure required to do the job and you'll most likely find it's only one or two key areas such as the seating areas and a sunny part of the lawn for sunbathing that you need to focus your attention (and cash) on, rather than the entire plot.

Once you've got an idea of your secluded spots (existing or planned for the future), sit on a chair and look towards any overlooking windows or views (sightlines). You'll immediately see that any screening elements that are brought in may not need to be nearly as high as you thought when standing. Use a tall stick and imagine it's a plant or structure and then ask someone to walk around the garden holding it in various positions while you're sitting down. You'll see that the closer it is to you, the more it'll block out any high-level views – from, say, overlooking windows or balconies – and often something around 1.5 to 1.75 metres placed right next to you will suffice. If you want privacy from direct next-door neighbours at ground level, it won't really matter where it is as long as it's between you and them and is above their eye level. This exercise will give you a clear idea of the height and width of the elements required and you can then work on what will work best within the style of your garden, whether it's a fixed structure such as a pergola or screen, or some simple planting or a combination of the two.

structures

Structures such as pergolas, gazebos or divisions set within the garden (fencing or trellis) are pretty instant and create an opportunity to grow some deliciously scented climbers, which work well near seating areas (honeysuckles, jasmines, roses). Large parasols, colonial shades and sail shades can work well too, especially when overlooked from above, but be careful with the wind, especially on roof terraces and take these down when not in use. Cost-effective willow, heather and bamboo screening are all available on rolls and can all be erected on posts (some are partly see-through, so check first). They tend to have a short life expectancy but are useful while plants establish themselves.

plants for privacy

Evergreen shrubs such as *viburnum tinus, laurus nobilis, pittosporum tobira* and *phillyrea latifolia* are good choices planted formally or informally and can also be clipped on the sides to keep plenty of height while limiting their spread. Clump-forming bamboos such as the *phyllostachys* and *fargesia* varieties can work well. Deciduous shrubs such as *cotinus, cornus, sambucus nigra, philadelphus* varieties and *euonymus alatus* all make good screening plants that will flower and provide seasonal interest too. They'll give you cover when in leaf in the summer for privacy but let increased light into the garden in winter when it's needed. You can buy large container-grown specimens at a premium that will give you instant cover. If the problem is being overlooked directly from above, look for trees with broad canopies that you can hide under. The hawthorn *Crataegus prunifolia* or crab apple *Malus floribunda* are good choices and will also add plenty of seasonal interest. If you live in a sheltered part of the world, then the evergreen loquat, *Eriobotrya japonica*, is an exotic addition with a broad canopy.

shade

I love planting for shade where lush tapestries and compositions can be created that burst into action in spring and cool the garden on a hot summer's day. Yet the gardening question I'm most frequently asked is how to deal with shade. Well, the very question already assumes shade is a problem/issue/negative. Let's rephrase it and pose the question 'How do I embrace shade and make the most of it?'. That's a more positive approach, isn't it?

All gardens have shady areas and some only have shade. It's worth categorising yours into light and dappled shade (perhaps filtered light coming through a tree's canopy), heavy shade and partial shade (areas that get some sun and some shade during the day). Then there's the level of moisture in the soil conditions at ground level from boggy to bone dry, with the latter being the most difficult to 'celebrate' plant-wise (note I didn't say 'deal with').

Turning deep shade into dappled shade

With solid structures like walls, fences and nearby buildings there's little one can do but if it's trees and larger shrubs that are casting the shade, perhaps their canopy can be pruned and thinned out to let more light through, which will increase the range of plants you can grow. Taking, say, one in every three branches out of a dense large shrub or small tree (which helps with air circulation too) and lifting the crown by removing some of the lower branches makes it stand prouder, look more elegant and change the light conditions beneath. With larger trees, call in a tree surgeon to advise you.

materials

I've designed and landscaped dark gloomy gardens that didn't receive any direct sun. Using tonally light materials for paving (light sandstone/tiles/light gravels) and painting or staining boundaries a light colour can make a huge difference by changing the gloomy feel of the garden and lifting the ambient light levels. Plants often respond too and it will increase the range one can grow.

planting in layers

The key to planting for shade is choosing plants that thrive in the conditions, so they always look healthy (avoiding those that may just survive but struggle!). I like to work in layers: a few shrubs for height and structure, perennials for a mid-story and ground cover to fill in and cover as much of the light-sapping dark soil on show. Losing that dark soil makes all the difference. You may not have room for a shrub layer or already have established shrubs to work with, which means you can just consider perennials and ground cover.

soil conditioning

This will significantly improve the success of shade planting, particularly in dry areas:

- Condition the soil well by adding plenty of organic matter such as garden compost, leaf mould or well-rotted manure.

- Plant more small plants as they have a better chance of establishing themselves than larger ones.

- Water in well initially and water when dry during the first year, especially during dry spells.

- Mulch, mulch and more mulch! Every spring and autumn, mulch to retain moisture and build up the volume of decent soil. If you cannot mulch the entire area, concentrate on each plant.

- As perennial and ground cover plants clump up, divide them (ideally in autumn) and fill in any gaps.

- When planting between tree roots, carefully dig a hole without damaging the tree roots to explore whether there is a decent depth for the proposed plant.

Some of my top plants for shade are included below.

Shrubs

- **Sarcococca confusa (sweet box):** A suckering plant that is very tough once established. Sweetly scented evergreen with white flowers from January to March.

- **Mahonia aquifolium 'Apollo' (mahonia):**
A tough, low-growing shrub about 1 metre tall, which spreads well. Spiny foliage turns a reddish purple in autumn and large clusters of yellow flowers in spring.

- **Viburnum plicatum 'Kilimanjaro sunrise':** The RHS plant of the decade! Pyramidal form with tiered branches and lacecap flowers in May and June. Best in dappled shade. Height 3 m x spread 2 m.

- **Pittosporum tobira 'Nanum' (Japanese mock orange):** Designers are turning to these evergreen shrubs for their low-growing structure.. Neat, glossy evergreen foliage. They grow well in protected gardens but may need some winter protection.

- **Hydrangea arborescens 'Annabelle' (hydrangea):** Hydrangeas do very well in partial shade if the soil retains moisture. This is a design classic, upright with white pom-pom flowers. Height and spread up to 2.5 m but can be pruned smaller.

- **Anemone japonica 'Honorine Jobert' (Japanese anemone):** Perfect for shady spots under trees and next to fences and walls and can cope with dry once established. Simple open, pure white single flower on tall wiry stems from around August onwards with large heart-shaped ground covering leaves. Height 1.2 m x spread approx. 60 cm.

- **Kirengeshoma palmata (yellow wax bells):** Large palmate leaves in spring and early summer that fade as the tall tubular yellow flowers appear. Give it plenty of depth and incorporate plenty of leaf mould. Height 1.2 m x spread 75 cm.

- **Angelica archangelica (angelica):** A statement plant reaching 2 metres in height, growing best in partial or full shade in a deep loamy wet soil. It has deeply cut leaves and rounded clusters of bright green flowers in summer. It's actually biennial so only lives two years, flowering in the second, but it freely self-seeds around. Height up to 2 m x spread 1.2 m.

- **Disporum longistylum 'Green giant' (disporum):** A fabulous perennial with a bamboo-like form that reaches 1.2 m. It has green-white dangling bell-shaped flowers followed by small black fruit. It is semi-evergreen in mild areas but cut the old growth to the ground in spring for fresh new shoots to emerge. Height 1.2 m x spread 1 m.

- **Hosta 'Devon green' (hosta):** Hostas are the shade lover's dream (if you can keep the snails off!). Plenty in shades of greens, greys and olive. This one has a cream edge to the heart-shaped leaves that will help light up a shady spot. Height 45 cm x spread 55 cm.

Ground cover

- **Brunnera macrophylla 'Jack Frost' (Siberian bugloss):** Deciduous ground-covering perennial with heart-shaped white leaves. Height 40 cm x spread 60 cm.

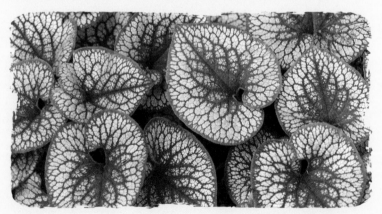

- *Vinca minor* **'Argenteovariegata' (lesser periwinkle):** Periwinkles are evergreen, carpet the ground well and grow anywhere. This one has a golden variegation to the edges of the sage green leaves and sprays of violet-blue flowers in spring. Clip back after flowering to keep tidy. Height 20 cm, spread 90 cm.

- **Asarum (wild ginger):** An extremely useful ground cover plant with glossy kidney-shaped leaves that reflect the light in the darkest of spaces. Height 15 cm x spread 30 cm.

- **Polystichum setiferum 'Herrenhausen' (soft shield fern):** There are hundreds of ferns for shade. Some are evergreen, some deciduous and there are those that like dry or moist shade. This one is evergreen and has fresh green filigree layered foliage. Height 50 cm x spread 90 cm. For very dry soils, try *Dryopteris filix-mas*.

- **Tiarella 'neon lights' (foam flower):** Masses of frothy, pinky white buds flower from May to July above deeply cut leaves with distinctive bronze veining. Carpets the ground well to suppress weeds. Height and spread 20 cm.

This may sound obvious, but the key to getting your garden storage under control is to have a good idea of what you want to store. I have seen hundreds of sheds and storage spaces full to the brim with all sorts of unwanted items (bikes, rusty old barbecue, bags of old compost, paint pots and the like). It becomes impossible to open the door, let alone get something you want out. Try to be organised, only keep essentials, and try to avoid filling a shed with the overspill of your interior belongings!

sheds

Conventional off-the-peg sheds are often the best value per square metre. Plan carefully where it will go, which way the door opens and where the window is, and make sure it goes on a decent base so air can circulate beneath (old slabs/concrete foundation with bearers or a modular heavy-duty plastic shed base – see below). The basic models can be easily personalised to suit your garden style (staining, cladding, covering with plants, etc.). There are also many ways to customise them to make them more practical by adding shelves, shelving units, hooks or large screws, clips, ties, rails, etc. so that every inch of wall and ceiling can be utilised. Do this before filling it up. Only fix into decent-sized timbers. A good trick is to up-spec the basic framework of a basic shed by screwing or bolting some 100 x 50 mm tanalised timbers and fixing them to an existing upright or horizontal beam or screw all the way through the walls into another piece of the same to form a grid or similar. A simple grid like this internally makes the shed stronger and now gives you options to fix screws into and hang some decent weight onto both inside and out.

Built-in storage

Building in storage can be a clever way to use otherwise dead space such as under exterior stairs or fire escapes or where space gets squeezed and inaccessible, say, between bay windows and a wall. These are often awkward places to get any other value from and may be better boxed in. Timber is by far the easiest material to use, but always use treated wood and exterior grade ply. Consider the fall of water at ground level and where that water will go (it may need diverting to avoid sitting under the unit).

Under-seat storage

Built in timber seats work well where space is limited but making bench storage attractive and completely watertight is not easy. I tend to build in storage with the intention of buying watertight storage boxes and then placing them inside. It will limit the size of objects you can put in them, but will ensure they stay dry. There are also some prefabricated free-standing bench units available that are both good value and (some are) watertight.

Bicycle stores

There are now quite a few off-the-peg bicycle stores available, which can be fitted into a front or back garden and locked for security. None of them are particularly attractive and they do eat up space. Consider making something bespoke or use wall brackets combined with a bike cover in a side alley or on a strong boundary wall. Our bikes are stored behind a wall and under a green roof made up of large, galvanised trays supported on a wall on one side and a couple of posts on the other, which looks neat and tidy and also means more space for plants in our small front garden.

Bin and recycling stores

Bins, especially large wheelie bins can be particularly unsightly but easy access is required for the collectors. They tend to disappear better when a store is custom built in a material and height that blends in with what's around, such as a gate, fencing or trellising. There are off-the-peg products with green roofs built in or make something yourself to fit (see the green roofs chapter).

Reclaim the ground lost! in this case, a spot for an alpine collection.

roof gardens and balconies

Roof garden considerations

There is something special about roof gardens; green spaces high in the sky above street level. As I walk around London, I find myself looking up to see if I can spot the tell-tale signs of one with foliage dangling tantalisingly over the edge of a wall or balustrade. Roof gardens can sometimes be retro-fitted onto old buildings, be integrated during the construction of an extension and many new-build properties include balcony spaces and larger terraces with reinforced concrete roofs that can take a decent weight. The approach to a roof space is always practicalities first; work out what can and what can't be achieved before getting your hopes up.

Weight loading and planning: Pots, compost and plants are heavy and if pots get blocked up and fill with water, they get even heavier. There's also the issue of how many people will be on the terrace plus any garden furniture. Planning permission, structural engineers, weight loading and how and where any structures (handrails and balustrades) can be fixed is paramount – you will most likely need professional advice in some way or another. Overload a roof and it can get dangerous and my advice is to never attempt to create what estate agents call 'an unofficial roof terrace'. I know of many that have been ordered to be taken down (time and plenty of money wasted).

Surfaces: There are some great products, including adjustable pedestals designed to 'float' tiles, decking structures and regular-sized paving slabs over the existing roof so water can freely drain through and no cement is needed so the top layer of the roof isn't affected or penetrated. A protective layer will need to be laid out on the roof first. Artificial grass has come on in leaps and bounds and now there's a range of looks to

choose from. It's easy to install – even as a temporary solution – and can look effective and fun while certainly greening the space up.

Containers and compost: Look for containers that are lightweight such as plastic, fibreclay, fibreglass, zinc or galvanised steel. Plan it carefully, measuring the space out and then think about how containers can be placed

An outdoor room in the sky!

for maximum planting. Consider having planters made up (wood or galvanised steel) to slot perfectly together with some taller and some shorter to create a tiered planting effect.

There are also some fabric planters (basically large lightweight bags) and even good old growbags that can supplement the main planting, fit in a few gaps and are ideal for a few salads, herbs or veggies. Plastic imitation terracotta and lead planters may sound a little naff, but some are thoroughly convincing and lightweight. If the weight loading allows, try to get at least some big planters too so you can add height into the planting. Use a drainage layer for good drainage to avoid waterlogging and use lightweight compost (some come ready-made for roof terraces and have light expanding clay aggregate (LECA) included in the mix).

Screening and windbreaks: Filtering wind with trellising and planting is far more effective than solid barriers, which are constantly buffeted and form damaging eddies on the garden side. For windbreaks, consider olive trees (surprisingly tough and hardy), Arbutus (strawberry tree), Escallonia, Phormium, *Viburnum tinus* and Griselinia.

Access: It may sound obvious but before you go shopping for plants, containers, furniture etc. make sure you know how you're going to get them up and measure the access before purchasing. I have seen elements that have neither fitted in the lift nor the stairwell being embarrassingly sent back! If it's a more substantial job or a long way up, a specialist contractor may advise you to get a winch or even a crane in, which is an involved process as it means coordinating deliveries like a military operation and arranging a licence to block the street but it can be done, at a cost.

Plants: Coastal plants can cope with strong winds. Silver birch, Tamarisk and *Cordyline australis* all do well and help with height but won't create much protection from the wind. Once you have created a buffer, plants such as dwarf pines, Abelia, Cotinus, *Phillyrea latifolia*, rosemary, lavenders, Phlomis and Cistus will help to build up a framework. Perennials such as Agapanthus, Centaurea, Crocosmia, pinks and pasque flowers all cope well with exposure and do well in containers as do ornamental grasses such as *Stipa gigantea* and the tall *Miscanthus sinensis* varieties. Any plants that don't do well in harsh winds or containers (bamboos, acers, etc.) should be avoided.

Irrigation: Regular watering is essential – containers can dry out in a day or two, especially where there's always wind. Although some people manage with long hoses dangling outside windows, if you can install an outdoor tap somewhere close then do. Drip irrigation kits and simple timers can be quickly and easily fitted, and these days are reliable, good value and often more efficient than hand watering.

Balcony considerations

Balconies require an identical approach to roof terraces. It doesn't matter how tiny the space is, to make best use of it, it's always best to have a plan to work to. Some simple measurements and a quick sketch will help you think about putting all the elements together (surface, pots, containers, seating, features), making the best use of the space and creating interesting views from inside. Ideally, draw it out on graph paper to scale. One of the best balconies I've seen (which was around 1.5 m x 3 m) had a fold-down table and chairs, a small wall-mounted water feature, lighting and fabulous planting!

calculating compost volume

Where space is tight, you won't want to have excess bags of compost lying around. A simple way to calculate a pot is to measure the surface area at the top (say the pot is square and the top is 0.5 m x 0.5 m = 0.25 m²). Multiply that by the depth, say 0.6 m, and we have the volume, so 0.25 x 0.6 = 0.15 m³. Multiply that by 1000 (litres in a cubic metre) and we now know it will take 150 litres to fill the pot (or roughly 2 x 80-litre bags). Hope that's clear!

front
gardens

Let us reclaim our front gardens! Over 5 million of them in the UK have no plants at all and there's been an increase of over 3 million completely paved over since 2005! Cars and car parking have had the largest impact, directly eating up the space, yet the knock-on environmental issues are enormous too: the loss of biodiversity (where we need it most in towns and cities) as well as increasing the 'heat island effect' created by hard surfaces. Our front gardens can bring communities together and affect our general wellbeing – we are all aware of walking down a street that is more green than grey.

Our family has the common problem of lack of space in the city: four bikes, a green waste wheelie bin, a general waste wheelie bin, a food waste bin and recycling bags a permanent fixture! It's a small terraced front garden but now the design functions well.

The bikes are locked outside and hidden with a green roof into which I've planted scented pinks, succulents, alpines, some spring bulbs and a few annuals. All the excess water that hits the paving drains back into the water table through the gravel under the bikes, the wheelie bins are accessible and less conspicuous as they are level with the top of the side fence, but there's still room for improvement and more plants; there always is.

considerations for front gardens

- Permeable surfaces such as gravel, pervious concrete setts (which let water through) or even grass sown into specially designed heavy-duty plastic modules (which means a car can drive on it) will let the water penetrate back into the water table rather than down a drain. This will help to reduce flash flooding.

● Climbing plants and wall shrubs take up little ground space, yet most will develop plenty of volume and can be grown up the front of the building, left to scramble over walls or even a ready-made structure such as a car port.

● Trees in a front garden always help to soften the building, tie it to the landscape and will also create a canopy over a parking spot. Many can be grown in a large container.

● Hedges rather than walls or fences filter dust pollution and can provide a home for wildlife too. If you have absolutely no soil whatsoever, then there's always the option to plant in containers and a few large containers will always be easier to maintain than lots of small ones.

- With planting, ensure a good proportion of evergreen planting as the garden will be seen daily throughout winter.

- Consider the colour and tones of the materials of the house and look to surfaces and walling that will tie them together.

- Consider the design from inside too. Look out of windows and think about what you want to see and views you may want to either obscure or create privacy from.

- Look to some scented winter-flowering plants (Sarcococca, Viburnum, Daphne, etc.) as they will never be missed and always be fully appreciated in a front garden.

- Consider pruning any overgrown existing or inherited shrubs or hedges. Crown lifting, thinning, and shaping may turn a messy plant into something majestic.

- With a symmetrical property where the front door is placed centrally, consider a formal layout and go with it with more formal planting.

- Use any existing features to your advantage such as tiering pots on steps or tying flower colours in with your front door.

- If you are planning on turning your front garden into a parking space, design it with room for plants. Avoid paving into corners so you can plant directly into the ground if possible. Think about where the car wheels will actually go. A couple of consolidated tracks for the wheels perhaps with low planting between and paths either side so you have enough room to get in and out can look great (the main expanse will be broken up with green, so it feels more garden than car park).

water features

Water introduces a cooling, intriguing and often playful presence to a garden space, especially in the summer. In towns and cities, the sound of moving water may not only be alluring in itself but will help distract from the hum of traffic or the sound of partying neighbours. Ponds are fabulous for wildlife but can be quite a project and not everyone has the space for one. Self-contained water features on the other hand take up little space, are far quicker and easier to install and smaller ones can even be put away for winter if they're not wanted. They have a reputation, perhaps conjuring up memories of low-budget TV makeovers, but it's what you choose, how you use them and integrate them into the garden that will ultimately make them work well. Some contain still reflective water, others can simply be placed, plugged in and turned on.

Types

There has been a bit of an explosion of water feature products in recent years. There is a huge range of styles  from traditional to ultra-contemporary. Keep your focus on your garden style (see city garden style chapter) and my advice is, keep it low key and simple. You tend to get what you pay for. Cheap ones may rust or degrade quickly and the pumps may pack in.

- **Mini pond in pots:** The simplest of all are sealed large pots or half barrels, which act like mini ponds and can sit on the terrace or in amongst planting. Some come with a small fountain but if you plant them up with some oxygenators, a marginal or two and perhaps a dwarf water lily, the water should be kept clear and it will hopefully  bring in nearby damselflies and dragonflies too. Small solar fountains can be fitted (note: waterlilies don't like moving water) for a little sound and movement.

- **Wall mounted:** If space is tight, you want to break up an expanse of wall or perhaps bring a focal point up towards eye level, then a wall-mounted feature  may be the way to go. They usually need screwing into the wall or sit on brackets. The biggest consideration is hiding the cable to the pump. Perhaps paint the cable the same colour as the wall, rake out some cement between bricks (so you can cleverly recess the cable into the joints?) and re-point. There's always evergreen ivy and other climbers that will grow over it.

- **Mini waterfalls:** These usually have a sequence of tiered cascading pools that fall from one to another. There are plenty of heavily themed products (Japanese bamboo, terracotta, Grecian, modern stainless steel, etc). These arrangements can work if you're looking to bring height into a space but, in my opinion, always look far better set in among plants rather than simply plonked on the patio.

- **Make your own kits:** You can buy a pump, sump and a few bags of pebbles that can be set into a raised area or dug into the ground. The idea is that the pump circulates the water through whatever you choose but drains back into the sump. You may have something (some rocks, a stone bowl, a ceramic evening class project!) you think would work and be unique for your garden and these kits can work well. Experiment a little, think about how the water can run over it or through it (perhaps by drilling a hole?) or dribble down it. Play around with it in situ. Make sure whatever you use is frost-proof and waterproof.

- **Silent features:** Some of the more contemporary water features are completely silent. Water runs over sculptural elements such as balls (they tend to be stone or stainless steel) or vertical walls (sheets of steel) without a sound. I guess they're clever, but I just find them slightly eerie and feel that if the water's moving, then why not be able to hear it? The sound is one of the best qualities, isn't it?

Siting and Planting

Water features' success relies on their siting and integration into the garden. I think they work best sited in the cool shadier parts of the garden set among lush plants such as ferns, hostas, large-leafed ligularias and rheum, which water is associated with perhaps more than sun-loving plants. It's always intriguing to be able to hear a water feature before seeing it as it triggers a feeling of wanting to find the source of the sound, so consider placing it just out of view or behind some plants, unless it's a particularly formal layout designed to draw the eye straight to the water feature.

Sound

For those with moving water, I can't stress how important it is to get the sound right. If you're not careful, water features can be extremely annoying and a distraction rather than an asset. They can sound either like a continuous peeing sound (no, not good!) or be way too loud and feel out of proportion to the space they're in. A trickle is good, a gush perhaps not. The sound will depend on the rate of flow, the height of the drop and the volume of water it may drop into. Buy one for which you can adjust the flow of the pump, but also play around with it once it's up and running. I've been known to jam hidden bits of cork into the spout to improve its sound.

Maintenance

All water features work by circulating the same water, usually with a submersible pump. If the water evaporates (as it inevitably does in summer), the most common problem is running the pump dry with no water, which is likely to break it. Keep the level topped up at all times and also change the water regularly to keep it nice and clean. Avoid using any soapy product unless you want bubbles in your water! Clean the pump properly at least once a year. Ideally, store the pump indoors for winter.

lighting

Designing with outdoor lights is fun and creative. It extends the gardens use and it will also make it a safer place in the evening (lighting steps, level changes, paths etc.). There are plenty of lighting techniques, but my advice is that as well as lighting key plants and features such as pots or any garden sculpture consider lighting some of the structural elements too. Pergola posts, boundary walls or fences, edges of paving, anything with a little texture that can help turn the summer garden into an atmospheric and different place in the evening. It can also look great from indoors too, creating a unique framed composition when viewed through a window and in the winter months when you get home from work you get to see your outside space rather than just a reflection of yourself in the window!

Mains lights

Mains lighting has the widest range of fittings, the brightest lights and is the most reliable. It is well worth considering if you are starting a garden from scratch but can be retrofitted, though cables will need to be dug into the earth or laid under paving. If it's a possible for the future it's always worth laying a conduit pipe under paved areas to give you an option to pull cables through it later, especially into planting areas that are cut off. A qualified electrician will need to install them, and the fittings can get quite expensive. Remember the aim is to subtly light the garden, not turn it into some kind of brightly light kitchen extension!

Solar

I love the way we can harness the sun's rays and use them to light our gardens in the evening. There's a large and ever-increasing range of solar lights available. They used to fizzle out early but with LED bulbs they can now pump out enough light for a whole evening. Post lights, spotlights, LED strip lights, deck lights, paving brick lights, moveable lanterns and torches as well as one-off party lights like nets and strings. Some have a small solar panel sitting on the top of the products themselves whereas others have moveable panels which you can place to discreetly catch the rays. Many are adaptable so the distance of the panel can be extended quite a way from the lights themselves, giving flexibility to light gloomy areas. I'd start off small and try a simple kit out in an area and see how they go.

Turns the garden into a different space at night...

candle lighting

Candle lighting is the cheapest and easiest way to light a space and can look great for a one-off party or when needing a little light to eat 'al fresco'. Tea lights placed in a jam jar or placed in brown paper bags with some sand in the bottom and placed on walls, edges of paths or tables are the easiest and cheapest way of lighting a garden. Play around, place them in lines or dot around rather randomly. There are many more fancy products with a huge range of lanterns and bowls which will help style the garden as well as provide a light source. You can also buy convincing battery-powered imitation flickering candles for outdoor use.

With all lighting be considerate to your neighbours and consider its effects on wildlife. Only use garden lighting occasionally and for a few hours.

Needs a bit of planning, but these LEDs look great and make the steps safe too!

garden lighting techniques

Think about using a range of techniques for variety.

Spotlighting: Spotlighting is used to highlight a particular plant (perhaps an architectural specimen or piece of topiary?) or feature such as a sculpture and ideally viewed from a reasonable distance.

Uplighting: Uplighting usually lights from ground level upwards onto say the interesting bark of a tree or into the crown of a large shrub. This technique can also be used on structures such as archways and pergolas to help keep height in the garden at night.

Grazing: Grazing is mainly used as a technique to light walls or hard landscaping structures in a garden. The lights can be placed at an angle to show off a structure or surface such as the bricks in a brick wall or the slats of a fence.

Downlighting and pool lighting: Downlighters set into a tree or fixed to a wall can be directed to create pools of light on specific areas of the garden such as a dining area or on certain features such as a path through an arch.

Underwater lighting: Waterproof lights can be placed under the water in ponds or fountains. These lights will give a glow to the water feature as a whole and emphasise any movement to the surface of the water. The underwater lighting of a waterfall or fountain can create a really dramatic effect and turn the waterfall into a cascade of light.

Silhouetting and shadowing: Shadowing is when a plant or sculpture is lit from the front to specifically cast a theatrical shadow onto a wall. The movement of plants like grasses and bamboos can also introduce subtle movement. The shadow will change according to the relative size of the object, distance between the light and its subject matter. Silhouetting is when the light is placed behind the object or plant and works particularly well with architectural plants.

green roofs

Green roofs are a great tool in the urban gardener's box and a fine way of growing plants where ground space has been lost through buildings. They have plenty of applications for the urban gardener such as flat roofs, shed and garage roofs, on top of garden studios, offices or even on top of a bike or bin store to help disguise the roof surface.

From a sustainability standpoint, they are valuable as they soak up rainwater to slow it down before it goes back into the water table or divert it from drains to help to reduce flash flooding and although they aren't proven insulators in winter (unless extremely high spec) they help to cool buildings in summer. Plants are also way better for biodiversity than a hard surface, but their biggest value probably lies in their aesthetics; looking out of a window onto a roof covered in plants is far more attractive and uplifting than a dull asphalt, zinc or felt surface.

There are different ways to build green roofs but with the introduction of some great products it's never been easier. The key is to know that the roof you're building onto can take the combined weight of the compost (when wet) and plants. With external buildings such as sheds, garages and garden offices, you can reinforce an existing roof by fixing some sheets of 20-mm ply covered with heavy-duty plastic sheeting over the entire existing roof, which will help spread the weight and extend its lifespan. Some bike and bin stores come prefabricated with a built-in green roof or you can make one with ply as above and fix an edging all the way round the outside to make what is in effect a large shallow planter. Before installing or retrofitting anything on roofs that are part of the house such as flat roofs on extensions it's advisable to get an engineer to check the weight loading, but if you are planning a new one, perhaps consider it as an integrated part of the specification?

There are some great new modular products on the market that make life easier. Plastic trays of approx. 500 mm x 500 mm (four to a square metre) clip together and can be placed directly onto the roof (simply roll out a landscape fabric beneath to place on). An edging material is then fixed to stop them blowing away in a gale. They can also be cut to fit and, once placed, filled with compost and planted up or seeded into directly. The chosen compost needs to be lightweight and free draining, so I'd suggest something along the lines of a 40% peat-free multipurpose compost, 30% sieved garden compost and 30% perlite or vermiculite. The latter are not trade names, but organic lightweight mediums made from naturally occurring volcanic rock to keep the soil nice and open. If you don't have any garden compost yourself then just make it all up with the peat-free bags.

As with all forms of gardening, picking the right plants for the right place is critical for their health and survival. Choose short stocky plants (no more than around 15 to 20 cm tall) that won't get blown around or suffer wind rock (which inhibits their roots rooting). Look for those that thrive in dry, well-drained soils and a blend of evergreens and deciduous plants: sedums, sempervivums and alpines such as armeria, aubretia, helianthemum, saxifrage, etc; creeping thyme and short grasses such as *Uncinia rubra* and festuca; small dianthus, such as *Dianthus deltoides* and *dianthus carthusianorum*; pulsatilla, the pasque flower, also does extremely well. In autumn, spring bulbs can be dotted in to extend the season.

The sedum matting cheat

You can buy sedum plant matting on a roll and pretty much place it on top of a roof like laying turf. It's not as exciting as a 'proper' green roof with a mix of plants but has many applications. There are technical instructions on websites about sub-bases and mats to lay underneath, but I have a large offcut at home thrown directly on my zinc garden office roof around 20 years ago and after many years of neglect, and drought in some years, it has never looked better. It's evergreen and flowers every year. You can also buy wildflower mats with a variety of species mixed together, but they are more expensive. For a roof, choose a plant mix that can cope with very dry conditions.

container gardening

Most small gardens will have at least some plants grown in pots and containers and some, such as roof terraces, balconies or areas with only hard surfaces, will most likely have all their plants grown in them. I often get asked which plants do well in a container and yes, some do better than others, but we generally buy most of our plants in plastic pots and with the right ongoing care (watering, feeding and potting on), trees, shrubs and topiary pieces can live long and happy lives in them with annuals, bulbs and some edibles marking the seasonal changes.

choosing containers

Containers can significantly help style a garden. The choice should enhance and go with the theme rather than look like an afterthought. A simple way to bring a city garden together and make it look more cohesive is to have all the containers within an area in the same finish and colour (metal, ceramic, stone, terracotta, fibreglass, wood). It will immediately make it feel more designed. Or you may want to go completely the other way and use a mix of containers, perhaps some recycled for an eclectic 'shabby chic' look, which can look great too. Anything that can hold compost and have drainage holes drilled into the base can be turned into a container. Metal containers such as galvanised steel or recycled olive oil tins can get

summer impact just where you need it!

hot (so bake and damage a plant's roots) if placed in the sun, so either line them with a thin sheet of polystyrene or place in shade.

Go big

In my opinion, a common mistake when gardening with containers is to have lots of small pots dotted around the space so you have to crane your neck downwards to see what's growing. Fewer large pots, perhaps planted with a combination of plants, is far more impactful and far easier to maintain too as small pots dry out quickly.

Tier them up?

Pots are ideal for placing at the base of retaining, boundary and house walls to help break up the elevation. They can also look terrific tiered up in groupings with tall pots towards the back and mid or shorter sized ones at the front; this way you'll see a mass of plants combining rather than the front sides of large pots. If it's a permanent placing, you may not see some of the pots at all, so to save money perhaps the obscured ones could be grown in plastic pots?

plants

Pots and containers are wonderfully versatile and often seen as a way of keeping the planting in a garden fresh and seasonal. Choose from annual flowers (such as cosmos, pelargoniums, snapdragons, petunias, nicotiana, *Orlaya grandiflora*, zinnia, marigolds, ageratum, verbena, fuchsia, arctotis, nigella and poached egg plant), spring and summer bulbs or edibles (herbs, dwarf French beans, peas, carrots, potatoes, salad leaves, tomatoes, strawberries, peppers, etc.). Most perennials and grasses look great in combos, perhaps for a seasonal display and then planted out into a more permanent spot when they've finished flowering. Structural plants like small trees, shrubs, exotics, and most climbers can also be grown in large containers.

compost and drainage

Drainage in containers is key and very few plants can cope with being waterlogged. Most outdoor containers have drainage holes already in them but it's worth checking before you buy as some pots are purely decorative and for placing plastic pots with plants within. I always add a drainage layer of 'crocs' (broken pot), gravel or broken up polystyrene packaging at the bottom to ensure good drainage.

What compost you use will depend on what plants you're growing.

Multipurpose compost: I only use peat free. This type of compost is suitable for most annual plants, bulbs and edibles.

Ericaceous compost: There are organic ones available for acid-loving plants like rhododendrons, camellias and azaleas.

Succulent compost: It's essential to get the planting medium right for succulents as they will suffer in a claggy, moisture-retentive compost or one that is too rich. Sharp drainage and low fertility are key, so something like a mixture of 70% John Innes No. 2 and 30% horticultural grit is ideal.

Tree, shrub and rose compost: A soil-based compost they can get their roots into such as John Innes No. 3 (or equivalent) with some added garden compost or well-rotted manure.

watering and feeding

- All plants grown in containers will require regular watering, especially in summer when they can dry out in a day.

- Most water is lost through plants' leaves rather than heat to the roots or soil, so if they're using a lot of water it usually means they're growing well!

- Use a finger check for moisture daily during warm or windy weather (even twice a day in hot weather!).

- If it has rained, it is still worth checking that the soil is moist, especially if the plant is placed next to a wall or under eaves.

- Water thoroughly. Fill to the rim. Let it drain into the compost, then top it up again to ensure that the whole of the compost is moist.

- Perhaps place a saucer or tray to capture the excess and save both water and nutrients.

- If water doesn't drain out of the bottom during watering, check the drainage holes for blockage.

- Grouping containers together to create a little shade will help reduce heat stress with container plants.

Feeding

Nutrients leach out of containers quickly and if a plant's roots are filling the pot, it makes it difficult for it to take in water and nutrients (see the Repotting container-grown plants chapter). During the growing season (spring to summer), feed your container-grown plants:

- If you buy compost in bags, it often has a feed in it, so check the label, but from April to the end of August you may need to use a general-purpose liquid feed or a high-nitrogen feed.

- Controlled slow-release fertilisers can be used and organic ones are available.

- After late summer, most plants stop feeding until mid-spring, although if short-lived annuals are still flowering they will benefit from feeding until early autumn.

- Feed when the compost is already a little moist.

- Liquid feeds, such as tomato feed and seaweed, are excellent for most flowering and edible plants grown in containers.

raised beds

Raised beds need an initial input of effort (yes, and of course some cash), but can be well worth it and perhaps the ideal solution for many urban gardeners. They can look great but also fundamentally change the way one gardens an area, significantly increase the range of plants that can be grown as well as helping to organise a space too, by introducing a more permanent structure. If you make some of them at around 45–50 cm height and wide enough, they can double up as space-saving garden seating.

From a gardening perspective, the main pros are the increase in drainage and being able to control soil and growing conditions better. Most herbs and salads love being grown in raised beds as well as a huge range of ornamental plants too and as it brings the working height up above ground level it makes sowing, planting and picking them easier too. They're also ideal for gardeners with mobility issues.

The material you choose depends on your budget and the aesthetics of your garden. Brick, rendered block and stone will all need foundations so are quite involved and become a permanent feature, so consider how they will complement the other garden materials? A less permanent choice and favourite is timber. There's no need for foundations or cement work, they're quick to install and easier to move in the future so less permanent. New railway sleepers are heavy, stack on top of each other and can be screwed together like giant Lego bricks so rarely need wooden posts to support them. Use new or nearly new ones as reclaimed old sleepers have often been soaked in creosote or oil and are not suitable for garden use (they have been classed as carcinogenic).

You certainly wouldn't want to grow edibles near them, and they seep sticky residues in the summer, which gets everywhere. New (oak or pine) sleepers are the way to go. They are approximately 200 or 250 mm wide so if two are stacked on each other on edge, it makes a bed 400 or 500 mm tall; an ideal height. A cheaper material is timber gravel boards or reclaimed scaffold boards, which will need the support of timber posts (driven or concreted into the ground) but are great on the allotment or for less permanent beds.

I always line the inside walls of a timber raised bed using some heavy-duty plastic sheet or landscape fabric pinned just below the top, so you don't see it. This keeps the soil away from the wood, stopping it from rotting and increasing the lifespan. I don't usually line the bottom unless there is a major weed problem (e.g mare's tail, bindweed). I would then use some landscape fabric on the bottom of the raised bed covering all the soil to create a barrier between the new soil and the ground. Water can get through the fabric, but the weeds will be suppressed.

If heavily compacted, fork over the soil beneath before backfilling. The soil and compost mix you choose to backfill with will depend on what you want to grow but for most plants I'd go for a mix of loamy soil and add in plenty of compost or well-rotted manure. Just using compost alone will be too free draining and won't retain enough moisture.

green walls

As we look to squeeze urban green into the tiniest plots, we view every wall or fence as a planting opportunity. To green up verticals, climbers should never be ruled out as a relatively low-maintenance solution, but green walls take it one step further. They can turn a wall into a visual feast with a range of plants; a little like having a garden border or green picture on the vertical. They're easily viewed at eye level and above and higher areas can be accessible too, so perhaps the ideal spot to grow some easy-to-pick edibles such as herbs or trailing strawberries? Small green walls are easy to make and with recent product development, products have never been easier, but the more ambitious large green walls are never easy. They can get prohibitively expensive, require technical specifications and regular maintenance, so become more of a horticultural art installation. There's nothing more depressing than a failed green wall where all the plants have died! My advice for green wall success is to plan it with precision and start small.

Modules and other products

There are an increasing number of off-the-peg products available, which is definitely the way to go. Some are plastic modules with built-in irrigation channels that slot together, while others are made up of recycled felt pockets (which stay moist) to place compost in and plant into. For a small green wall patch, some products are ready-to-go, while others can be added to and scaled up into something bigger.

Suitable walls and weight

The combined weight of the modules, compost, plants and water, when watered, is significant. Only fix to a sound strong wall and use fixings recommended for the product. Moisture and damp is another consideration, especially on house walls. The backs of most of the products trap moisture in, so only fix to house, boundary, garage or shed walls that you know are completely sound already.

Watering and feeding

Before venturing into installing a green wall of any description, consider how the plants will be watered. A south-facing sunny wall may need to be watered every day in summer to keep the plants alive. They are usually watered from the top and gravity feeds the water through the system. Consider what happens at the bottom too when the (brown) excess water drips out. Will it splash onto paving or go down a drain? Ideally, there will be a planting area beneath to soak it up and benefit. The best way to feed any plants during the growing season is with a liquid feed (weak seaweed or tomato feed mixes are best for the majority of plants) that can be added into the irrigation water every few weeks or as required.

Plants

It's surprising how many plants it takes to plant up a green wall – more than you'd use on the ground – so factor this in with any calculations. The planting pockets aren't big, so buying small pots (such as 9 cm) is the way to go and they'll quickly fill the space. It's key to plant accordingly for the aspect. A green wall is often drier and sunnier on top and wetter and shadier lower down, so look to plant accordingly.

Some shady side ornamental plants to try: perennials – ferns, bergenias, ajuga, heuchera, tiarella, vinca minor, pachysandra, epimediums, *Cornus canadensis*, campanula, viola, wallflowers.

Sunny-side ornamental plants: geum, scabious, salvias, helianthemum, *Erigeron karvinskianus*, sedum, pelargoniums, petunias, lobelia, bacopa.

Herbs and edibles

You can plant seed directly into green wall pockets (some edibles like radish and beets don't like being transplanted) or pop in some plants from small pots or plug plants. Most edibles require sun. Herbs (thyme, marjoram, mint, basil, sage, etc.), salad leaves, tumbling tomatoes, radishes, Asian greens, chard, and many others can all do well in green walls. Pick regularly to keep compact.

trees for
city gardens

Every garden, no matter how small, should have at least one tree. When I go into a garden without one, I instinctively sense there's something missing. Perhaps it's the connection to the larger natural world but from a design perspective they play an important role of adding height and form to balance the overall planting scheme. Deciduous trees mark key seasonal changes with their flowers and autumnal colours, and some have magnificent bark and stems.

City dwellers are fearful of planting trees in small gardens, worried they will grow too big and so dominate and create shade, or be sited too close to the house and over time the roots will affect the foundations, leading to structural problems. Sure, we need to be sensible here and not plant trees like oaks, eucalyptus and the dreaded cypress leylandii in small gardens, but there are many fine choices out there. As a general rule, make sure that you plant it at least its ultimate height away from the house.

Height and spreads are maximum, and each can be thinned and reduced if they're getting too big.

Trees for small gardens

Amelanchier lamarckii (snowy mespilus): Fine as a single stem or more sculptural multi-stemmed form. White star-shaped flowers in early spring with bronze leaves maturing to dark green and then turning a fabulous fiery orange and red in autumn. Eventual height up to 10 m x spread 10 m.

Olea europea (olive): Grey-green leaves and tiny, creamy-white, summer flowers. This elegant, evergreen is hardier than most think and makes an excellent specimen plant for a sunny spot in the ground or a large pot. I've grown it successfully on roof terraces where it copes well with strong winds. Height 5 m x spread 5 m.

Betula (birch): Birches work beautifully in the city alongside strong architecture. The striking white bark of a silver birch such as the elegant *Betula utilis* var. *jacquemontii* 'Grayswood Ghost' is a fine choice with its buttery autumn colour and yellow catkins in spring. Height up to 18 metres by 10 metres spread. There are many other birches too such as *B. albosinensis* 'China Rose' with gorgeous, huggable red-pink bark (height 10 m x spread 8 m).

Sorbus (rowan/mountain ash): I adore most ornamental ash trees as they reflect every season with flowers, autumn colour and berries. *Sorbus vilmorinii* has creamy white flowers in spring that are followed by clusters of berries that fade  from red to white as the season progresses. The feathery foliage turns a dark crimson in autumn. What more could you want! Height 5 m x spread 5 m.

Magnolia: If you have soil on the acidic side, there are a wide range of wonderful magnolias for the small garden. The evergreen *Magnolia grandiflora* will add formality and can be grown as a bush or standard tree, while the deciduous M. 'Star Wars' is a light pink, long flowering new hybrid, which will stay small at 3 m. The yellow flowered *M. denudata* 'Yellow river' has large goblet-shaped creamy yellow flowers in April/May and slowly reaches 10 m.

Malus (crab apple): The ornamental crab apples are ideal for a naturalistic look, encouraging wildlife and plenty of seasonal interest. *Malus* x *floribunda* has pale pink flowers in mid- and late spring, followed by small, golden-yellow fruit often persisting into winter, which the birds love to feed on. Height and spread approximately 8 m.

Cercis siliquastrum (Judas tree): Clusters of pink pea-shaped blooms burst from the new young shoots to cover the tree in colour before the leaves come out. From late summer onwards, large bunches of rich purple pods deck the branches and last well into winter, while the pretty foliage turns light yellow and mahogany in autumn. Height 10 m x spread 9 m.

***Acer* (maple):** Japanese maples can work extremely well, adding a touch of zen to a small city garden. Most thrive in shady conditions too as long as there's enough moisture in the soil. The Japanese maples such as *Acer palmatum* 'Sango-kaku' and 'Osakazuki' both reach about 6 metres and have intense autumn colour.

***Prunus* (cherry/plum/peach):** Ornamental and fruiting cherries will provide a stunning display in spring. One of my favourites, however, is the more delicate winter-flowering *Prunus* x *subhirtella* 'Autumnalis', which flowers when much needed from late autumn to early spring when the garden is otherwise fairly dormant. Height 7 m x spread 7 m.

Arbutus unedo (strawberry tree): This strawberry tree has glossy evergreen foliage and a wonderful shredding cinnamon-coloured trunk. Its white flowers are dainty white bells, and the fruit is similar to strawberries. Edible? Yes. Tasty? No! Perfect specimen tree and good for large containers. Height up to 8 m.

***Cornus kousa var. chinensis* (Chinese dogwood):** Okay,
technically a large shrub but designers love it for its
change in seasonal interest. Fabulous long-lasting white
showy bracts (like large flowers) in spring. Yellow, purple
and crimson autumn colour and strawberry-like red fruit.
Can be grown on neutral soils but will colour up best on
a more acidic soil. Height 7 m x spread 5 m.

shrubs

Shrubs form a year-round woody structure above ground and are deciduous or evergreen. Some get as large as small trees and they play the role of the mid layer in a planting scheme. Many have multiple seasons of interest. Here are some of my favourites for city plots.

Shrubs for spring and summer flowering

Buddleja davidii 'Royal Red' (the butterfly bush): Its deep reddish-purple flowers reach 20 cm long, have a strong scent and are great for butterflies and pollinating insects. Most *B. davidii* varieties reach around 3 to 5 metres in height if left. Prune hard back in April to a pair of strong buds. If you prune to about 60–90 cm from the ground, they should grow to about 1.8–2.4 m so flower where you can see them.

Hydrangea paniculata 'Wim's Red': There are so many great hydrangeas that love dappled shade. Great value plants that flower for ages in high summer, their blooms change colour throughout the season. This one reaches around 2 m with a strong upright habit and its honey-scented cone-shaped blooms change from white to pink and then a deep wine red. What's not to love!

Philadelphus **'Virginal' (mock orange):** Lovely vase-shaped form with masses of highly perfumed white flowers in June and July. Height around 3 m. When established, prune out one in every four branches to the ground after flowering to keep an open and balanced shape.

Sambucus nigra **'Black Beauty':** I don't normally go for dark-foliaged plants as they can get lost in some gardens but when in flower this is a stunner with its deep purple, almost black, lacy foliage that set off the sprays of white flowers to perfection. Cut to almost ground level every year to promote the best foliage.

Roses: Hundreds of roses to choose from (shrub, climbing, rambling, ground cover, etc.) My simple approach is to pick a colour and flower shape, and then make sure it's repeat flowering and scented! Modern roses are generally disease-free so that's a big factor. Shrub roses are the easiest to grow and maintain. 'Claire Austin' (white and gorgeous), 'Munstead Wood' (velvety crimson) and Roald Dahl (orange-red buds open to a soft apricot) are three of my faves.

shrubs for great autumn colour

Vaccinium corymbosum (blueberry): As well as producing delicious fruit in summer, blueberry plants turn an intense deep crimson in the autumn, making them one of the best shrubs for autumn too. They prefer acidic soils but grow well in a container using a compost mix of ericaceous compost with some added leaf mould mixed in. Water only with rainwater. Height 1.5 m x spread 1 m.

Cotinus obovatus (smoke bush): Many cotinus have purple leaves in summer but I prefer this more natural looking plant with its green, rounded leaves. In autumn

it changes its character with fiery red, orange and purple foliage that looks spectacular against its dark stems. Height approx. 6 m x spread 6 m. Can be pruned hard back in early spring if so required.

Ceratostigma willmottianum (Chinese/hardy plumbago): Low grower that carries fine autumn foliage colours moving from yellow through purple into red but also has divided mid-blue flowers from late summer into autumn. Versatile plant that I use a lot for softening edges of harsh paving and tricky corners, which it fills nicely. Height 1 m x spread 1.5 m.

Euonymus alatus (winged spindle): A dense shrub with mid to dark-green leaves in summer and the most fabulous strawberry-red autumn colour. Reaches 2.5 m

in height – but the smaller 'Compactus' is 1.5 m – but I often clip it into a nicely rounded shape in spring to keep it even more compact and add form to a small garden.

Enkianthus perulatus (enkianthus): *E. campanulatus* is quite a large shrub, reaching around 4.5 m in height and spread, whereas E. 'perulatus' is smaller at around 1.5m in both height and spread so may be more suitable. Dainty white downward nodding bell-shaped flowers in spring and then the most extraordinary autumn colours of orange and red as if someone's lit it with a match and stood back. Best on acidic soils.

shrubs for winter scent

A seasonal highlight is the assortment of plants designed by nature to pump out their sweet winter perfume in order to lure in the few pollinators around at this time of year. In a city garden, you may not have room for that many, but place those you have where you're bound to get a good whiff, so, close to the back door or in the front garden where you're guaranteed a daily encounter.

Elaeagnus ebbingei (oleaster): Tough cookie coping well with wind, shade and dry soils once established. Dark glossy evergreen leaves with contrasting silvery undersides that are revealed in a breeze. The autumnal flowers are small but highly fragrant with tones of lemon, ginger and coconut. A friend once said they smelled of Malibu (the drink) and he was right! Height and spread 3–4 m, but can be clipped to size and makes a fine hedge.

***Mahonia x media* 'Charity' (lily of the valley bush):**
Another tough plant that will grow in any aspect or soil
(except waterlogged). Flowers early from November
onwards and will keep going right through Christmas.
Deeply cut architectural leaves and upright spikes of
bright yellow sweetly fragrant flowers. Eventual height
5 m x 4 m spread but easily kept to 2 m x 2 m if pruned
hard after flowering.

***Sarcococca hookeriana* var. *digyna* 'Purple Stem' (sweet
box or Christmas box):** Sarcococcas are essential scented
winter evergreens flowering from December to March.
The new stems on this one have a dark purple tinge to
them, and the delicate sweetly fragrant flowers have an

 attractive pink
base. It forms
a nice stocky
plant. Height
1.5 m x 2 m
spread but
can be pruned
back after
flowering.

Viburnum farreri (viburnum): Upright deciduous shrub with bronze leaves in spring that turn a dark green in summer and then purple again in autumn. Pink-white highly fragrant flower clusters appear from November to February as the leaves drop and the flowers persist on bare stems. They may be small but don't half pack a punch. Height 3 m x 2.5 m spread.

Daphne mezereum (daphne): A hardy early daphne flowering around February. Violet-pink flowers appear in dense clusters along the bare stems and are followed by shiny red berries in summer. Strong, sweet fragrance. Shady woodland conditions. Height and spread approx. 1.5 m.

evergreen shrubs for clipping, shaping or topiary

All gardens benefit from year-round structure in the form of clipped evergreen plants to hold them together. They don't have to be formal and can perhaps be more organic in shapes and sizes.

Box: *Buxus sempervirens* is a classic but there are many others and some good alternatives, as the dreaded box blight is an increasing issue. *Ilex crenata* (boxed leaved holly), *Phillyrea angustifolia* (narrow-leaved mock privet), *Taxus baccata* (yew), *Prunus lusitanica* 'myrtifolia' (narrow-leafed Portuguese laurel) and *Osmanthus x burkwoodii* are worth considering.

perennials

Perennial plants push up in spring with a wonderful freshness and then reliably flower. Some bloom for long periods and then naturally die back in the autumn, their roots building up energy for the following spring. Many look great as they die back in the winter (so don't cut back) and are beneficial to wildlife. Most are clump-forming so can be split and popped into gaps, drifted through borders, or planted in blocks to create impact. They can be grown and combined nicely in pots too. There are hundreds of perennial plants but here are some stalwarts that always do well and look great in city gardens.

perennials for small gardens

Anemone x hybrida **'Honorine Jobert':** Japanese anemones are tough, long flowering and rarely need staking. This one has pure white single flowers with a pink tinge on the underside of the petals. Light green and yellow centres. Height 1.2 m x 0.6 m spread (approx. three plants per square metre).

Astrantia major **'Claret':** Astrantias bloom well in sun or shade and have a long flowering period in high summer. This one's got deep ruby-red flowers, dark stems and deeply lobed leaves. Use lighter (white and light pink) varieties in shade to stand out.

Cenolophium denudatum **(Baltic parsley):** Brings a touch of the wild to a small garden. Tall upright umbel with creamy white flowers above ferny foliage. Ideal for dappled shade. Height 1 m. Five plants per square metre.

***Persicaria amplexicaulis* 'Firetail' (red bistort):** Height 1.2 m x 1.2 m spread. A vigorous plant with slender leaves and spikes of crimson red flowers between July and October. Makes a dense carpet so ideal for covering ground but adds plenty of height when in flower. Semi-evergreen so may keep its leaves in mild areas. One plant per square metre.

***Eurybia x herveyi* (new name for aster):** An elegant plant smothered with lavender blue starry flowers with deep yellow centres. Ideal for sun, dappled or partial shade. Height 1 m x 0.5 m spread. Approx. three plants per square metre.

Geranium **'Rozanne' (hardy geranium):** Geraniums are essential city plants, tough, shade tolerant and flower for ages. Many cross over into the ground cover department too. 'Rozanne' is the RHS plant of the century! Large, open, mid-blue disc-shaped flowers with white centres that flowers pretty much continuously from spring through to early autumn. Height 60 cm x 75 cm spread.

***Nepeta racemosa* 'Walker's Low' (catnip):** Aromatic sage-green foliage and violet-blue flowers from June onwards. Loves sun but copes well with partial shade. Great for edges of paths or spilling over a small wall or raised bed. Height 60 cm.

***Salvia guaranitica* 'Amistad':** Salvias come in a huge range of sizes and colours, most flower for ages and in my view are essential garden plants. They need sun and good drainage. 'Amistad' is a winner with deep purple flowers and dramatic black hoods and can flower from May to November! Height approx. 1 m.

Gillenia trifoliata (**Bowman's root**): Masses of small star-shaped flowers and good autumn colour. Light and delicate looking plant for partial shade but requires some moisture in the ground. Height 1 m.

***Verbena bonariensis* (purple top):** I rarely leave this out of a garden as it takes up so little ground space and offers so much. Tightly packed purple flowers cluster at around 1.8 m on wiry stems and rarely need staking. Flowers from June to September and beyond and butterflies and bees love it. Height 1.8 m

ground cover

Some gardeners leave huge gaps between their plants, but soil is one of the worst mediums to visually set off most plants, being dark, dull and matt. It saps any available light and too much on view only increases the drab gloomy feel of a shady urban space. Unplanted soil is also poor for the soil structure as nutrients quickly leach out of it and it provides areas for weeds to get a grip too, so increases the overall maintenance of a plot. See these empty patches as an opportunity to get more planting into your garden. Ground cover plants (many are woodland plants coping well with shade) do just that, knitting together quickly to cover the bare earth and visually set off the taller plants. Try to plant in quantity (more than one or two!) to form a textural carpet beneath the taller plants.

Ground cover plants for small gardens

***Euphorbia amygdaloides* var. *robbiae* (wood spurge):** Native evergreen wood spurge and the first to flower. Zingy lime-green long-lasting flower-like bracts that make a fabulous foil for others including bright coloured tulips. Rosettes of dark shiny foliage last all year. Can cope with dry shade once established. Height 60 cm in flower by approx. 80 cm spread.

***Ajuga reptans* (bugle):** Short spikes of deep blue flowers and shiny evergreen leaves that knit together nicely making it an extremely useful shady low ground cover choice. Versatile too, doing well in paving gaps, to soften edges and even in low pots. Bees love it. Height 15 cm x 50 cm spread. *A. reptans* 'Braunherz' has burgundy leaves and Ajuga 'Catlin's Giant' has oversized burgundy bronze leaves.

Epimedium x warleyense **(Bishop's hat):** There are many *epimediums* out there to choose from. This is perhaps the showiest with eye-catching burnt orange flowers in April and mottled heart-shaped foliage all year round that have a good autumn colour. A tip for this deciduous plant: cut back any leaves from last year to show off the flowers nicely and make room for the new leaves to come through. Height 50 cm x 60 cm spread.

Dicentra **'Stuart Boothman' (bleeding heart):** Gem of a spring flowering plant with thinly cut grey-green foliage and nodding deep pink flowers on arching stems. 'Bacchanal' is another good choice with deep-red flowers. The fine foliage stays on into summer. Prefers a slightly alkaline soil. Height 30 cm x 40 cm spread.

Vinca minor 'Bowles' Variety' (lesser periwinkle):
Periwinkles are as tough as old boots and make an
excellent evergreen ground cover in shade and can cope
with quite dry conditions once established. This one has
light blue flowers and dark glossy green foliage. A lovely
white variety is *Vinca minor* 'Gertrude Jekyll'. Will send out
shoots to spread but you can snip back after flowering to
keep it upright and tidy. Height 20 cm x 50 cm spread.

Brunnera macrophylla 'Looking Glass' (Siberian bugloss):
This is one of the best ground cover plants for a gloomy
spot. The heart-shaped white dusted, veined leaves form
a neat dome and help to bounce any available light

around. Delicate looking sprays of bright blue flowers in spring. Height 35 cm x 45 cm spread.

Pulmonaria **'Blue Ensign' (lungwort):** There are many lungworts, all of which are tough plants to choose from. 'Blue Ensign' has bright violet-blue flowers from March onwards and dark-green foliage. P. 'Sissinghurst White' has white flowers and more typical white spots on the leaves. Height 35 cm x 45 cm spread.

***Asarum europaeum* (wild ginger):** A designer's favourite for a shady spot and grows well in most soils. Glossy, rounded, kidney-shaped leaves and unusual hooded purple flowers appear in spring. Low growing to about 15 cm.

***Stachys byzantina* 'Silver Carpet':** An excellent plant for a sunny, well-drained spot. Soft, hairy silver leaves are ideal for breaking up the edge of paving or the edge of a raised planter. Purple flowers mid to late September.

***Thymus* 'Silver Posie':** Thymes are deliciously scented, evergreen and will carpet the ground well if given sun and sharp drainage. Ideal for planting between stepping stones or in cracks in paving. Cut back after flowering to keep compact. This one has white margins to the small leaves and purple-pink flowers in spring.

***Bergenia* 'Overture' (elephant's ears):** Bergenias will grow pretty much anywhere (sun, shade, dry or moist soil) and their mounds of large glossy leaves form an excellent texture when planted in informal quantity or edging a border. The flowers range from whites, through pinks to purples. 'Overture' has wonderful deep-red leaves in the winter and bright magenta flowers in spring.

***Polystichum munitum* (sword fern):** Evergreen ferns really earn their keep and can combine well with other ground cover plants either planting in numbers or as mini eruptions through lower ground cover plants to play with form and texture.

climbers

I'm always looking for climbing plants and combinations of climbers that will help cover fences, walls and trellises for every aspect. They are also useful to grow up divisions in the garden, soften pergolas and gazebo structures and scramble up obelisks and pyramids as a feature in their own right. Climbers help break up the garden visually at eye level and above, and the smaller the garden gets, the more important they become. Evergreen and deciduous climbers as well as wall shrubs should all be considered as options but don't forget annual climbers too, which will help fill a gap (while more permanent plants establish themselves?) and flower all summer long for minimal outlay.

climber support

- Make sure to give them the support they need. Some will self-cling, some like to grow on wires, but plants with smaller tendrils such as clematis and passion flowers may prefer a smaller mesh to scramble up.

- Prepare the planting hole well with plenty of moisture retentive organic matter, especially next to walls, which dry out the soil.

- Plan their placing when putting up a new fence so that you plant in the gaps between posts and don't try to plant into the concrete footings.

- Consider maintenance levels. Some climbers are far better behaved than others and will need less tying in and pruning. Think how you will do these jobs and how you can easily access them without trampling on others.

- Don't always go for a quick, fast growing solution. Some climbers such as 'Russian Vine' will quickly become a nuisance and almost impossible to maintain.

- Plan generous planting pockets into paved areas adjacent to walls and fences so climbers can be grown into the ground rather than in pots as an afterthought.

climbers for small gardens

Akebia quinata (chocolate vine): Semi-evergreen (will drop its leaves in a cold winter or certain areas) with exotic leaves. Its early ruby flowers dangle like earrings and carry a slight vanilla fragrance with a hint of chocolate. Quite rampant so show it who's boss every year after flowering. May produce sausage-shaped fruits if there's another close by. *A. quinata* 'Cream Flowered' has creamy flowers with maroon centres. Height 10 m x 2 m spread.

Jasmines: Intoxicating summer perfume. *Jasminum officinale* (the common white jasmine) is a fine choice if you have plenty of room to let it do its thing and does best in a nice sunny protected spot out of the wind. *J. officinale* 'Inverleith' has the added attraction of pinky-red buds. *Jasminum polyanthum* (sweetly scented jasmine) is widely regarded as a house or conservatory plant too and the one you see grown on a hoop in the florist's. They're not reliably hardy but can do well for years outdoors in a protected spot. Their scent is superb and early in the year too. The star jasmine (*Trachelospermum jasminoides*) isn't actually a jasmine but I'm including it as it does carry a classic jasmine scent. Designers love them, being evergreen, neatly clothed right down to the ground with glossy leaves. Most labels say 'full sun', but I've grown it in dappled and semi-shade in a sheltered spot and it has flowered its socks off for weeks.

Ivy: Much maligned, but forms a fine evergreen backdrop and great for wildlife. Most garden varieties are relatives of English ivy, *Hedera helix*, many ideal for the city garden. *Hedera helix* 'White Wonder' is a variegated form that can significantly lift a shady north wall. *H. helix* 'Ivalace' is slow growing and compact with deep green and crinkly edged glossy leaves and *H. helix* 'Chicago' has a classic ivy leaf, dark green with grey veining.

Wisteria: Majestic climbers for a sunny spot. Far more reliable than they used to be – sometimes people waited years for a flower that never showed up! Try to buy grafted wisterias (look down the stem for a graft point) in bud or flower, as it's a sure sign it will flower again. Some good ones include: *Wisteria sinensis* (Chinese wisteria) with lilac-blue flowers and the most powerful fragrance of them all; *Wisteria floribunda* 'Rosea' with pink flowers and 'Alba' with long white pendulous flowers reaching 60 cm; and *Wisteria floribunda* 'Multijuga' (Japanese wisteria), one of the showiest wisterias around with extremely long flowers up to 1 metre.

Clematis: There is pretty much a clematis for every month and they're so varied; some delicate, others full-on blousy. The spring flowering montana varieties and evergreen *Clematis armandii* the most rampant and many (*C. viticella* and *C. jackmanii* varieties and others) can be grown up into a climbing rose or tree.

Climbing roses: As with all roses, a mind-bending selection is available. Scented, repeat flowering and disease-free is my motto. 'A Shropshire Lad' (dark pink buds that open to softer pink). 'New Dawn' is a tidy grower and hugely popular and smothered in quite small creamy pink flowers. 'Souvenir du Dr Jarmain' is not a 'new rose' but has gorgeous deep crimson flowers and almost thornless stems and is good for a shady north-facing wall.

Virginia creeper: Deciduous and self-clinging, ideal for hiding an eyesore! Some get huge so I recommend *Parthenocissus henryana*, which has prominent white veins in its leaves and turns a fine autumn colour. May need cutting hard back occasionally.

perfect for that shady wall!

***Hydrangea anomala subsp. petiolaris* (climbing hydrangea):**
One to mention as it does well on a shady wall, is
self-clinging and produces large lacy white flowers in
summer. Quite slow to establish itself but worth the wait.

***Passiflora* (passion flower):** Exotic-looking plants (quite a range to choose from) with large intricate flowers in summer sometimes followed by orange fruit. Pretty vigorous and more resilient when mature but is not fully hardy and may need some winter protection.

Holboellia: Exotic number. Twining with glossy leathery leaves and scented flowers followed by unusual purple-white sausage-shaped edible fruit. Needs a sheltered wall and wires or trellis to scramble up. *Holboellia latifolia* is commonly available.

annuals

Here comes the summer!

Annual plants grow, flower (for ages) and set seed all in one year. They can be grown from seed or bought in small plugs or pots. They are mostly tender plants (so wait till after the last frost before planting out) and most only last a season before composting. Annuals are great for packing colour into beds and pots, filling gaps or as a temporary solution for young gardens while other plants establish themselves and are great to play around and experiment with.

Annual flowers

Cosmos: One of the classiest annuals around and work well anywhere. Huge range. 'Purity' is white with yellow centres, 'Rubenza' is deep plummy pink and short and stocky so rarely needs staking. *Cosmos atrosanguineus* 'Chocolate' is smaller and more delicate, ideal for pots with velvety deep-red flowers that smell of pure cocoa.

Rudbeckia (black-eyed Susan): Cheery daisy flowers in a range of yellows and oranges with darker mahogany tones. 'Chim Chiminee' looks like a sweep's brush, 'Cappuccino' is darker, and velvety. They're both around 60 cm, whereas 'Gloriosa Daisies' is taller at around 90 cm.

Sunflowers: Best in borders. They're always such fun and make great cut flowers. Do you go for the biggest, 'Russian Giant', to see who can grow the tallest or something more in scale with the border like 'Velvet Queen' with its coppery red petals? Don't deadhead these as they won't keep producing; let it seed to attract birds to feed in the autumn.

Cleome **(spider flowers):** Cleomes are elegant and statuesque plants with unusual flowers held high (about 1.2 m) on strong stems so ideal for the middle of a border. 'Helen Campbell' has pure white flowers and 'Colour Mix' is a jolly range of whites, deep pinks and purples all together.

Zinnia: Get your sunglasses out! Zinnias are like miniature dahlias (between 15 and 75 cm tall) and thrive in a hot dry spot. 'Aztec sunset improved' has tones of yellows, reds and oranges (all in one flower!), 'Miss Willmott' is lipstick pink and 'Zahara Starlight Rose' has white flowers with deep pink centres.

Geraniums: Now we call these annuals geraniums but they're actually Pelargoniums, spot on for a Mediterranean look in window boxes or pots lining a set of steps. Available in reds, whites, pinks and some oranges. Trailing varieties are perfect to break the front edge of a pot or hanging basket.

Snapdragons (*Antirrhinum*): Kids love them of course and they remind us all of our childhoods. You can buy in single colours like the pure white 'Royal Bride', or some mixed-up crazy colours like 'Circus Clowns'. 'Bronze Dragon' has pink flowers and deep purple, almost black foliage to set them off.

Petunias: If you want your window boxes and hanging baskets to compete with a pub garden display then *Petunia surfinia* classic is what you need: trailing plants ideal for baskets and tubs in mixed colours of red, white and blue! 'Trillion Bells Celebration Mix' is more compact and coordinated, with pastel shades, or just go for a single colour.

Tobacco plant (*Nicotiana*): Strong perfume and an elegant form. Mainly whites, pinks, purples and greens. 'Marshmallow' is a mix of pastels. *Nicotiana sylvestris* is my favourite, tall and majestic (1.2 m) with cascading scented white flowers that pump out a heady scent on a summer's evening.

White laceflower (*Orlaya grandiflora*): This is a very stylish plant with white umbels of white flowers with tiny sails at the end of the petals. Great for hoverflies. No choice either, just the white one! 60 cm tall. Looks great planted en masse in a border.

Annual climbers

Ideal for quickly covering fences and walls or push some canes into a border for them to scramble up.

***Ipomoea* 'Heavenly Blue' (morning glory - not to be confused with the weed!):** Large, fresh, sky-blue flower (10 cm across) with white centres that unfurl in the sun. Each one only lasts a day, but they just keep coming all summer. Any well-drained soil will work if kept moist and in full sun. 'Purple Haze' has intense violet purple flowers. Height 3 m.

Sow seeds indoors in early spring, then stand back and watch!

Ipomoea lobata (**Spanish Flag**): A vigorous climber with unusual and exotic-looking flowers produced in a small rack-like effect along one side of a long stem. They are orange-red at one end and fade through to a cream at the other. The foliage turns a purple tint later in the year too. Full sun in moist, well-drained soil. Height up to 5 m.

Asarina barclayana (**Angel's Trumpet**): A naturally twining climber that grows to the ideal height of around 1.6–2.2 m, perfect for walls and fences in partial shade. It has rosy pink snapdragon-like flowers that dangle downwards. Technically a tender perennial so can be overwintered indoors but easily grown from seed as an annual climber. Height approx. 2 m.

Cobaea scandens: The cup-and-saucer plant develops a woody frame in its native frost-free tropical Central and South America and may make it through a very mild winter, but generally we usually grow it here as an annual climber. It's still quite rampant and has lush foliage and large cup-shaped purple flowers and prominent stamens. *Cobaea scandens* 'Alba' has white flowers. Height 3 m.

***Thunbergia alata* 'Suzie Hybrids' (black-eyed Susan):** Bright golden-yellow flowers with dark centres adorn this rampant climber reaching 2.5 metres. In mild areas it may make it through the winter outdoors. It enjoys heat, sun and plenty of moisture/humidity so hose it down when dry.

exotics

In the summer, exotics gardens come into their own. Stepping outside the back door can feel as if you've just walked off a plane into another country. This style works particularly well in small city and town spaces; the planting can be condensed, impact instantly created, and the chosen plants tend to thrive and perform in the microclimate offered. Some specimens can put on ridiculous amount of growth; my banana plant has been known to shoot up over a foot in one day – drama itself! This style is quite easy to create but a little trickier to keep going as some plants will need special care and overwintering (see overwintering exotics). I'd suggest going for a few easy and hardy exotics to set the scene (bamboos, phormiums, chusan palms) and then add in perennial plants such as ferns, hostas, asarum, euphorbias, grasses and persicarias to fill in beneath and generally increase the verdant feel. Colourful crocosmias, day lilies, salvias and dahlias are easy choices to bring in shots of bright colour in summer.

Cannas: Fabulous plants with huge leaves and, if the summer's warm and long enough, showy flowers too. Many to choose from: *Canna* 'Durban' has orange flowers and dark, purple-red leaves with pink veins; C. 'King Humbert' has bronze leaves with deep-red flowers. Height up to 2 m.

Fabulous foliage!

Hedychium (Ginger Lilies): Perennial plants that, with a deep winter mulch, can come back year on year. The Khali ginger *Hedychium gardnerianum* has stout stems with fleshy leaves and, up above, tall spires of deliciously scented yellow and orange flowers. *H. greenii* is reliable, has dark-green leaves with a maroon underside and orange-red flowers.

Yuccas: Okay, some are spiky numbers so they come with a safety warning (keep away from paths and children) but their evergreen, hugely architectural forms are hard to beat. Some yuccas such as *Yucca gloriosa* and *Y. filamentosa* 'Bright Edge' are surprisingly hardy, make ideal statement plants and flower sporadically with tall spires of creamy white flowers. They need full sun and sharp drainage.

Bananas: The hardiest and most reliable banana is *Musa basjoo*, which has huge leaves up to 2 m long. Grow it out of the wind in sun or shade and if the trunk is protected through winter, it'll develop plenty of height, otherwise it will reliably re-shoot from the base. Mine had an amazing flower followed by (small and inedible) bananas last year.

Astelia: Excellent silver-leafed foliage plants that can be grown in partial shade, so help lift a planting scheme. *Astelia* 'Silver Spear' is the daddy with dusted sword-like leaves up to 2 m. It's commonly available but look out for A. nervosa 'Westland', which is smaller at around 40 cm and has coppery tones mixed in and makes a great ground cover. Give it plenty of moisture.

most silver leaves love sun, but this does well in shade too.

Melianthus major (honey bush): A very easy perennial with a spectacular leaf form. Large, arching, deeply cut and serrated-edged grey-green leaves. It's a bit of a beast reaching up to 3 m in height but that's what we like! Weird dark-red almost black flowers appear after a long hot summer. Mulch with plenty of dry straw to get it through the winter.

***Tetrapanax papyrifer* 'Rex':** It used to be hard to get hold of but fortunately now is far more commonly available. Unmistakeable huge palmate leaves with a gingery underside, just one will set the exotic scene. If given winter protection, it will develop into a fabulous large shrub; if not, it'll behave more like a perennial and come up year on year.

Dicksonia antarctica **(soft tree fern):** This is the most commonly available, hardiest (to −10°C) tree fern in the UK. They have strong thick trunks with gingery brown hairs and fresh green fronds that reach 2 m or more. Very slow growing up to 4 m so buy (invest in!) the size you want.

Love a tree fern! It can cope with the deepest shade.

Hardy palms: There are a few palms that survive outdoors in most areas of the UK: *Trachycarpus fortunei* (Chusan palm) with a classic palm silhouette and hairy trunk; *Cordyline australis* (Torbay or cabbage palm) which is not technically a palm, but looks like one; *Chamaerops humilis* (Mediterranean fan palm), a suckering plant with masses of fan shaped leaves; and *Phoenix canariensis* (the Canary Island date palm) with its huge feathery leaves and solid trunk.

spring bulbs

Planting a range of spring-flowering bulbs is one of the most exciting times in the gardening calendar. It's like painting a picture onto your garden with invisible ink, something to get ridiculously excited about until, in spring, all is revealed. If you plan a good range such as narcissi, tulips, crocus, scilla, camassia, iris and alliums, you could have bulbs flowering from January through to May and sometimes beyond. For impact, think big. Large swathes of colour if possible, rather than little fireworks going off here and there. Bulbs are versatile and can be grown in borders, lawns and containers and compared to a bunch of cut flowers or even a cup of coffee they are great value too.

Bulbs for late winter/early spring include: snowdrops (plant 'in the green' in spring – still with leaves on); Chionodoxa (glory of snow), rock garden iris, crocus, early daffs, grape hyacinths and anemone blanda. Mid-spring, try: hyacinths, scilla, mid-daffodils, Erythronium, fritillaries, Dutch iris and Leucojum. In late spring there are alliums, Nectaroscordum, Camassia...

Some bulbs are one-hit wonders and won't flower again (most tulips), but most will come back year on year and naturalise by setting seed and clumping up to form colonies over time.

How to plant bulbs

- Orientate the bulbs with the growing point facing up.

- As a general rule, plant at least three times the depth of the bulb, with a similar distance between individual bulbs (except in pots where you can really pack them in).

- On heavy and wet soils, drop a handful of sharp sand into the hole before planting to avoid them rotting off through the winter.

- If you're planting a lot of bulbs or planting in lawn areas, consider buying a bulb planter, which takes out a neat core of soil that is easily popped back on the top.

- For the naturalistic look, put the bulbs in a bucket and toss them in the area you want to plant and plant them right there. Resist the temptation to move them around!

Bulbs in pots

- Use a multipurpose compost (ideally peat-free).

- Place a drainage layer over the hole at the bottom of the pot.

- Fill some compost and then nuzzle the bulbs in a little deeper than they would go in the ground.

- Pack them in closer than you would in the ground even if it means ignoring the packet's instructions.

- Add compost and finish with a layer of grit or gravel as a decorative finish.

- If you have a squirrel problem (digging up the bulbs) the only way to combat them is with chicken wire/wire mesh pinned over the top.

plant snowdrops 'in the green' (when the bulb still has leaves after flowering in spring). plant all other spring bulbs in autumn.

Bulb lasagne

If you have some left over or buy some bargain bags, consider planting a 'bulb lasagne' (I'd like to stake a claim on the term having come up with it on *Gardeners' World* around 15 years ago!). Get a large pot and cram bulbs into it layering them on top of each other, packing compost in between. Works best with around three layers. Put the largest, such as tulips and hyacinths, in first as they go in deepest and flower last and then add in some Narcissi followed by smaller bulbs such as Muscari, Crocus or *Iris reticulata*. Once planted up, you could even place it in a really shady spot to help brighten it up as the energy to grow and flower is already stored in the bulb.

practical
gardening

General Planting Rules

- Prepare the soil by removing weeds and stones and dig in some organic matter (homemade compost, well-rotted manure, mushroom compost, etc.).

- When planting a few or a border, always place all plants first to get an idea of spacing and visualise how they'll grow together.

- Make sure the plants are moist, perhaps soak them in a bucket for an hour or so until bubbles stop coming up.

- Dig a hole with a spade or hand trowel to roughly the depth of the plant's pot.

- Remove the plant carefully from the pot. Gently tease out its roots (especially if they're going round in circles).

- Place the plant in the hole and adjust the level at the bottom by adding or taking away so that the surrounding soil level will be the same as it was in the pot. This is important: not too shallow or it'll expose roots and too deep will bury the crown.

- Backfill with excavated soil and firm in around the plant with your hands so the plant's roots are in contact with soil and there are no air gaps. For larger plants and trees, lightly tread with your boot but don't squash all the air out.

- Water it in thoroughly (even if the soil is moist) to ensure the roots' contact with the soil.

watering

Try to plant sustainably so your plants won't need watering all the time. Most trees, shrubs and perennials will only need watering while they establish themselves (perhaps the first year or two) and overwatering can lead to waterlogging or weak root structure.

- Water either early in the morning or in the evening so that the water doesn't evaporate during the heat of the day. The morning is preferable as softer plants may develop fungal diseases if left wet overnight.

- Avoid wetting delicate or young leaves in direct sunshine as they may scorch.

- Light sprinkling is a big no-no, you're often better not watering at all. If only the soil surface is wetted, the plant's roots are drawn up to the top, discouraging strong root development and making it less likely to cope with dry periods.

- Water thoroughly and right down into the soil or compost around the roots, rather than the plant's leaves. A hose with a nozzle or a gun on the end means you can control it and stop the flow instantly, leading to less wastage.

- Scratch the surface of the soil because although it may be dry on top, there may be plenty of moisture an inch or two below.

- Stand pots in saucers to prevent water draining away and place them out of the midday sun so they don't dry out so quickly.

- Lawns may go yellow in a dry summer without water, but will quickly green up when it rains.

Mulching

Dressing the top of the soil around plants with any well-rotted organic matter locks moisture into the soil, reduces weeding and improves the soil over time.

- Never mulch when the soil is dry as it will exacerbate the issue by drawing moisture out, so wait till after a heavy rainfall or water key plants before mulching.

- Avoid building up mulch around a shrub's stem or covering the centre of a perennial or any small delicate plants, which may encourage them to rot off. If you're not sure you have enough material for an entire area, mulch around each plant individually and then fill in the gaps.

- Mulch with a minimum of 5 cm, and ideally 7.5 cm, of organic matter on flower beds.

- Organic mulches include: wood chippings, chipped bark, well-rotted manure, leaf mould, spent mushroom compost (alkaline), seaweed, spent hops or homemade compost from the compost heap.

- A square metre at 5 cm deep will require 50 litres of material. Many come in manageable 25 and 50-litre bags, but with larger areas, and if you have easy access, look at getting a bulk loose delivery, which works out far cheaper.

pruning

Yes, I know pruning can strike the fear of God into some! Just remember the three Ds (dead, diseased and damaged) and you won't go too wrong. Always use clean, sharp tools. As a rule, prune lightly after flowering (only if necessary) and any major pruning of larger branches should be done in the dormant period between autumn and spring. Sure, you may lose some flowers the next year but that's the way it goes!

- Always use sharp, clean tools: secateurs, loppers and a pruning saw (an ideal hand tool for woody stems – some cut on the push and pull stroke and others on the pull). With larger branches, a bow saw may be needed.

- Remove dead, diseased or rubbing branches first. Then, look to shape the plant to balance it out.

- Snip off any frost-damaged tips back to the next healthy bud.

- Remove excess growth where it's too dense so that air can circulate between the stems. This reduces the chance of fungal diseases and allows beneficial insects access to munch on pests.

- Always cut back to just above a bud (not into it) and, if there is a single bud, then this should be outward facing so that the plant develops an open shape.

- Cut at an angle so the slope is away from the bud so that water runs off it. This also helps the plant's hormones be directed into the bud, triggering growth.

composting, wormeries and leaf mould

Compost is one of those things that keen gardeners just love to make. We love the process (composting completes the cycle in the garden and the seasons) and adore the results. Raw unusable stuff goes in and extremely useful stuff comes out. Rather than seeing shredded letters, potato peelings and old newspapers as rubbish or stuff for the recycling box, see them as perfect compost fodder. Compost itself is an important source of nutrients for healthy plants and if added regularly and in enough quantity, will significantly improve the structure and fertility of your garden soil. Sieve it, put it into pots and you can sow seeds directly into it. Any gardener looking to avoid using chemical feeds relies on this magical stuff and the resultant healthy plants are far more likely to be able to stave off pests and diseases, so chemicals can be avoided once again.

Many now have a 'green waste' recycling bin in their front garden, a fine way to get rid of green waste, but it's hard to get the resulting produce back. Home composting is more like having a small distillery in your own back yard. It will ensure you get all the produce from your own material and you know precisely what's gone into it. Although a pile of garden waste left in the corner of a garden will naturally break down into compost over time, a structure made for this purpose will generate more heat and be easier to manage, making it more efficient all round.

which compost heap to make or buy

They're easy to build out of old wooden pallets simply nailed together or you can buy one off the peg in kit form. Space will be at a premium so site your compost heap where it doesn't take up valuable growing space.

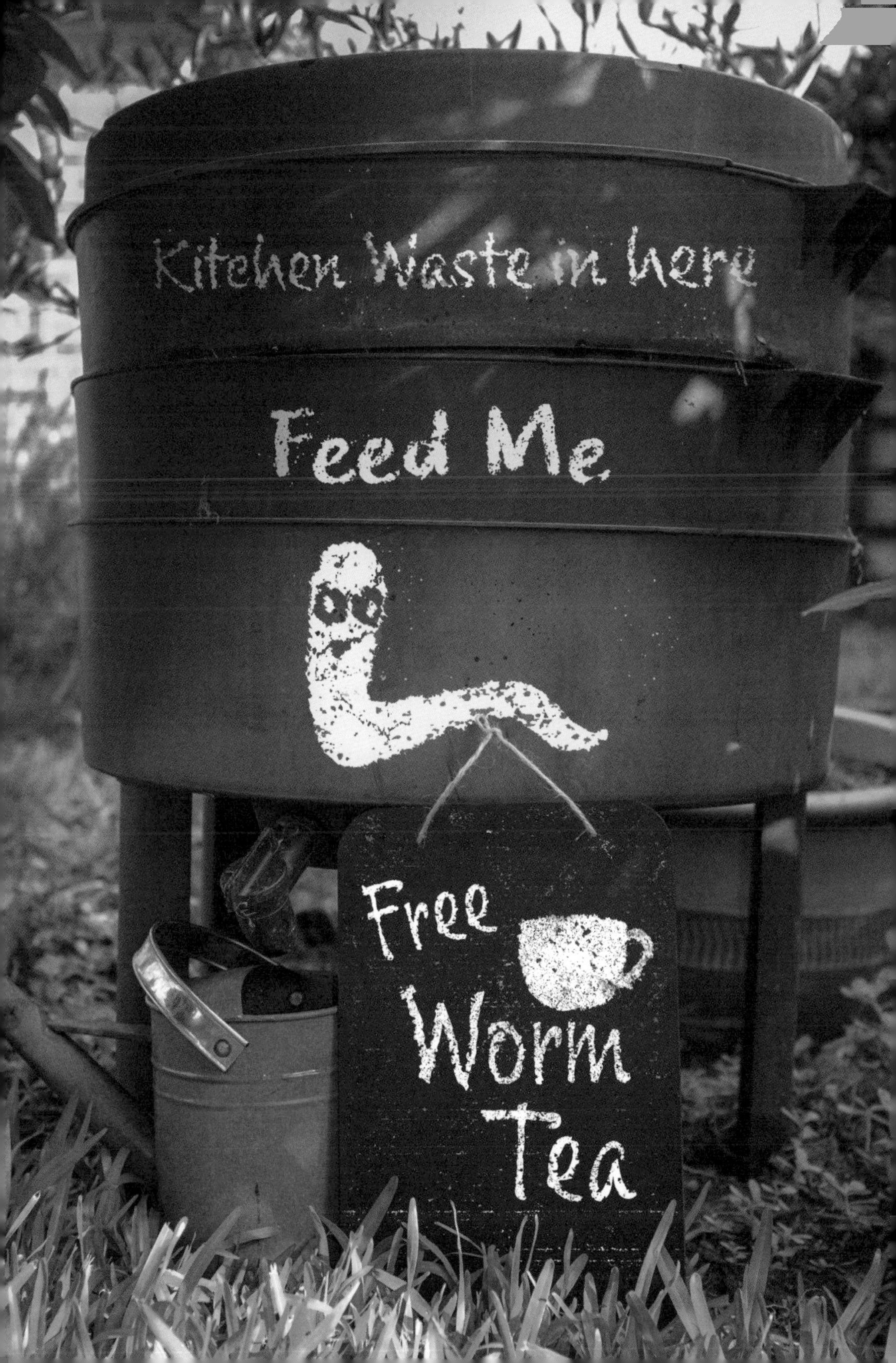
Kitchen Waste in here
Feed Me
Free Worm Tea

The bigger the better, but of course the urban gardener doesn't have much space. Many different designs are available from rolling barrel types (space-saving and effective for small quantities) to fancy ones with different compartments. If you have room, a no-nonsense homemade pallet design is a classic, the ideal size of about 1 metre all round. Clad the outside with cardboard to retain heat, which will also break down in time.

what to put in

The key is to get the balance right. Too much of a single waste will delay the decay and result in an unbalanced end product. You'll quickly become instinctive about what it needs. The best proportion is 50/50 of 'green' and 'brown' materials mixed together, which is perhaps trickier to find in city houses and gardens.

Greens: vegetable peelings, leaf material from plants, grass clippings (in layers and not too much at once), nettles, annual weeds, comfrey, tea bags, coffee grounds.

Browns: cardboard, cardboard tubes, shredded paper, hay, straw, sawdust, wood shavings, hedge clippings, fallen leaves (again in layers). Other good materials include eggshells, 100% natural clothes fibres and some wood ash in moderation.

what not to put into your compost heap

Do not put in any plants with disease or virus (signs are mottled leaves) or any pernicious perennial weed roots or seed/seedheads from weeds. Councils compost at a high temperature that kills these off, but on a domestic level it will never get hot enough.

moisture and turning

When the heap is full, make sure the contents are moist by sprinkling some water on top and then cover it with some sacking or plastic sheeting. It's best to turn the heap, say twice a year, to get some air into it and accelerate decay. Keen gardeners with space have two or more heaps so they can empty it from one to another and use some half-decayed material to start off a new heap. You can tell when it's ready as it's a consistent black and crumbly medium, which only smells earthy.

Pick out any twigs before using if they slipped in by mistake and apply liberally. Oh, and when your recycling collectors ask you why your bins for garden waste and recyclables are empty, tell them you've decided to keep hold of all the valuable stuff yourself!

wormeries and worm bins

Some city gardeners have a wormery and not a compost heap and some have both although the former isn't a substitute for a compost heap. Wormeries take up less space and turn kitchen waste and small amounts of garden waste into compost and also produce a concentrated liquid fertiliser which is great for diluting (10:1) and feeding plants with. Wormeries have two compartments, one for the composting and one as a sump for collecting the liquid. Composting worms can be bought or will turn up of their own accord and are different to earthworms. I had one for many years and it worked well although it doesn't get rid of much waste throughout a year and in truth getting a wormery to run smoothly is trickier as the working worms need to be looked after (moisture levels, avoiding heat in summer and sub-zero temperatures in winter, etc.).

See leaf raking as the start of something...

Leaf mould

You may view raking up fallen autumn leaves as a bit of a chore. Some left on the ground, perhaps in a patch in a wilder area is certainly beneficial for overwintering critters. Wet leaves smothering small plants may encourage fungal diseases and left on the lawn they can turn it yellow and encourage difficult fungal diseases too. They can clag up ponds, making the water super smelly and if left on paths and paving, can turn them slick and slippery. That's the garden health warning over.

If you see the leaves as the starting point of something more horticulturally valuable then the whole task may feel more fulfilling and Zen. I know someone who runs out in the streets with bin bags to grab them before the council sweeps them up! Leaf mould is extremely easy to make, and is simpler than general garden compost as it doesn't need a mix of ingredients, just leaves.

Which leaves?

Deciduous leaves such as oak, beech and hornbeam are the best and most commonly used, leading to Grade A leaf mould when sieved, but smaller ornamental trees such as acer, apple or cherry are fine too; in fact, leaves from any deciduous tree are fine. Evergreens such as aucuba or laurel hedge clippings are best chopped up first and added in thin layers where they'll break up quicker and turn into a general compost. Conifer needles are best kept separate and bagged up alone as they take longer to break down and form an acidic mulch, which is ideal for your azaleas and rhododendrons.

The containers you make it in will depend on how many leaves you have and how much space you have to store.

The bin bag method

Pack leaves into large perforated black bin liners as you collect them. If they are dry, dampen them with some water first. Tie the necks of the bags, stab them with a fork to let in a little air and any excess water out and stack somewhere out of the way while they rot down.

The wire cage method

Make a leaf mould cage. Hammer four 150 cm long wooden posts into the ground to form a square measuring about a metre in each direction. Tie wire netting round all four sides using garden wire. Throw the fallen leaves into the cage as you collect them and then dampen thoroughly. Tread down well and cover the top of the heap with a sheet of plastic (this speeds up the rotting time and keeps the moisture in). Keep adding new layers till the supply of leaves runs out then ideally cap the heap with a 5-cm layer of soil. If it's slow to break down, try turning (emptying it and filling it up in reverse) to speed it up.

Uses of leaf mould

High-quality sieved leaf mould is excellent for seed sowing, you can make your own potting mix (mix with loam and grit), whereas general leaf mould makes a good all-round soil improver when dug in or used as a mulch. It aids moisture retention and improves drainage so if added year on year will help transform a poor soil into a very decent one. A few buckets added to your compost will help boost its breakdown rate too as the gazillions of magical microorganisms get to work.

repotting
container-
grown plants

Pot and container-grown plants need to be treated differently to those grown in the ground. They need regular and increased watering as they come into growth in spring (and feeding too). Nutrients in the compost are either taken up by the plant or leached out through the pot so without feeding they can struggle to grow bigger let alone produce fruit or flowers, and over time become weak and prone to disease.

All container-grown plants require repotting every two or three years. If left, they will become 'pot-bound', which is when their roots completely fill the pot and even start going round in circles. The pot can't retain moisture as there are more roots than compost and nutrients can't be drawn in by the fibrous and usually diminishing 'feeding roots'.

when to repot

Spring is the perfect time to repot most plants as they'll come into strong root growth immediately after being potted. Check if a plant is pot-bound by turning it out of its pot and looking at its roots. If you can't get it out of the pot or its roots are coming through the drainage holes at the bottom, it's a sure sign it is pot-bound! Black, smelly and/or weak roots signify it's waterlogged so think about improving the drainage when repotting. Even if it isn't pot-bound, the compost over time will lose its texture and ability to hold moisture and most probably become overloaded with deposits from water and feeds, another reason why every two to three years is about right.

Definitely time to repot!

size of containers

It's sometimes tempting to pot up into a much bigger pot, but this means that the plant will only rock in situ (which leaves air gaps between roots and compost), or the compost can become very dry or waterlogged – it's difficult to get it right. One size up is ideal, allowing one or two inches all the way round. With large trees and shrubs, you can go a little bigger. Yes, you do end up needing a range of sizes of pots but if you can repot a few plants at the same time, perhaps you can cleverly step them up from one to another, always cleaning pots thoroughly before reusing. Small pots dry out really quickly so consider combining some plants together in a larger container to look better and reduce maintenance.

Drainage

There is a new school of thought that does away with a drainage layer altogether, but I still like to make sure there's absolutely no chance of waterlogging. A 2.5 to 5 cm layer of crocks (bits of broken pots) or broken polystyrene packaging (much lighter) over the container's bottom and drainage holes is ideal.

compost

See the container gardening chapter.

Root pruning and top pruning

With container-grown trees and shrubs, they'll eventually become impossible to repot into larger containers, which means getting serious (don't be scared, you can do it) and making room by pruning off around a quarter to a third of the roots and potting back into the same container with fresh compost. It will in effect arrest the plant's development a little like a bonsai would. Scrape away the compost to expose the roots' structure and always use a sharp clean pair of secateurs. Prune back the roots (take damaged and weak ones off first) and try to leave as many of the smaller fibrous feeding roots on as possible. Take off some of the top growth at the same time to keep the plant compact while balancing out the root-to-leaf proportion, which should avoid stressing the plant out too much.

TIPS

Water: Water plants thoroughly a day or two before repotting as their roots are then less liable to dry out during the process.

Water-retaining granules: Consider adding into the compost for more moisture-loving plants (sparingly as instructions) as they swell and hold moisture and reduce watering.

Tricky plants to repot: Plants like Agapanthus actually like to be quite pot-bound as it mimics their natural habitat of growing between trees' roots in shallow soil and there are also containers that are rounded with narrower openings at the top. The plants will eventually have to come out but are stuck. One way to get them out is to cut a 'V' into the middle of the clumping root with a sharp knife or old serrated bread knife, take out the middle section, which should create room to manoeuvre to get the rest of the plant out without breaking the pot.

Turn back to front: Plants such as mint are tough and grow very quickly and may need repotting every year. A good way to keep them the same size is to take the plant out of the pot, cut the rootball in half, turn those two halves back-to-back (so two curves meet), trim off any excess overlapping the pot edges, plus a bit more, and place back in the pot. You will now have some gaps to fill with some freely draining compost and the plant will fill the pot again in a season.

Top dress: In years when you aren't repotting, scrape back some of the old compost from the top (around 5 cm if possible) and cover with new compost.

protecting
exotics
in winter

Many town and city dwellers exploit their protected microclimate to grow at least a few exotics. Plants like hardy palms, agaves, bananas, tree ferns, oleanders, olives and cannas (see the Exotics chapter) look fab in small gardens, transforming a space into an urban jungle.

Winters have become increasingly unpredictable, but it's always best to play it safe with these kinds of plants you've invested in. Be well prepared and have a plan in place before temperatures plummet in autumn.

Move if possible

If a plant is grown in a pot and can be moved by lifting (bend your knees!) or wheeling around on a sack trolley, then move it into a frost-free conservatory or greenhouse if you're lucky enough to have one. The conservatory needs to be cool, not centrally heated, as the warm dry air and change in temperature will probably end up shocking it and doing more damage than good. If you don't have a greenhouse or conservatory, place them together in the most protected spot in your garden. This could be down a side alley, a basement well, in the corner of the garden where two walls meet or even under a large, preferably evergreen tree as the air will be marginally warmer where they'll be protected from damaging cold winds and less prone to ground frosts. Consider buying or making a temporary greenhouse/ polytunnel type structure out of bent plastic pipe or a wooden frame clad with plastic sheeting or bubble wrap or even a lean-to tent affair against a wall. If you use plastic, make sure plenty of air can circulate around each plant. Most of these types of plants despise winter wet more than anything and will cope with colder temperatures if they are dry, especially in the case of plants from arid conditions such as agaves, yuccas and echeverias. Watering will depend on what you're growing but reduce watering significantly, don't water if freezing temperatures are forecast and dry-loving plants may not require watering at all. Place pots off the ground (using bricks or pot feet) so they don't freeze and ensure good drainage.

Grown in the ground

If your tender plants are grown directly in the ground, you'll need to do whatever you can in situ. Unfortunately, whatever you do may be rather unsightly as you'll probably end up creating strange ghost-like figures in your garden. In milder areas, wrap plants with winter fleece or readymade fleece bags or jackets and secure with string or wire as they can easily be blown off. Old hessian sacking works well too for larger stouter plants like tree ferns – just make sure they can breathe. For colder areas, construct a simple framework around the plant using chicken wire and stuff it with plenty of dry straw for insulation. This works particularly well with palms, tree ferns and bananas, the latter of which will, in turn, develop a tall trunk rather than dying back into the ground like a regular herbaceous perennial. The older a plant gets, the tougher it'll become. The cabbage palm, Cordyline australis is a pretty rugged character, but may benefit from having its leaves tied into a bunch above its head like a disco diva, which will keep the worst of the frosts and wet out of the most vulnerable area, the crown, from where new growth appears in spring.

Don't use plastic or bubble wrap directly around plants as it will simply hold moisture in, make them sweat and considerably increase their chances of rotting off.

Cannas and gingers are a different story as they die down into the ground. These are a bit trickier as the advice varies depending on where you live. In milder areas you can leave them in the ground, cut them back to 6 inches above ground level as soon as the foliage blackens in the frost and then mulch thickly with at least 10 cm of organic matter. Plants left in the ground may

not flower so well next year as they come into leaf late and therefore have a shorter time to bloom, but it's far less hassle than digging them up to overwinter. If you live in a cold area, you have no choice but to cut the foliage off, put the rootball into a plastic bag with a few holes in it, or a large pot, and put it into a frost-free garage or greenhouse and make sure it doesn't dry out. It'll sit there dormant but should be fine for planting out the following spring and hopefully come into flower late summer.

tooled up

Buying the right tools for the job will not only make life easier and save you money in the long run but will also have an effect on storage too. What tools you'll need will depend on which type of garden you have; if you don't have a lawn, you won't need a lawnmower! Buying cheap tools is a false economy whereas good quality tools are more efficient, can last a lifetime, and are far more comfortable and sometimes a joy to use.

Basic set

Spade and fork: I go for stainless steel forks and spades with wooden handles as they tend to be lighter, cut through the soil easily and wet clay sticks less. They are easy to keep clean and handles can be replaced if broken. Check for a balanced weight and handle you are comfortable with before buying, choose either regular (larger for forking over and digging larger patches) or smaller 'border' forks and spades (big enough for most city plots and more nimble to use) or perhaps both depending on the job. Keep the spade sharp with a sharpening stone, which makes digging and cutting through the odd root much easier.

Rake: There are different forms depending on your needs. A lightweight leaf rake made from heavy-duty plastic is best for leaves on lawns, soil or even paving surfaces, whereas a wire or spring-tine rake can do the same job but is also for lawn scarification. A soil rake is only used to level and refine soil areas, which may not be an ongoing job in a small garden.

keep your spade
nice and sharp!

Trowel and hand fork: Used for hand weeding and planting small plants and bulbs. There's a wide range of shapes and sizes available but look for a pair you feel comfortable with. I particularly like a sharp, long, thin-bladed hand trowel to dig out deep roots of perennial weeds and plant bulbs in between other plants.

Watering can: Metal is more durable, but heavier than plastic. Look for the largest you can easily carry when full to save you having to go back and forth. A free-flowing and easy-to-pour 'rose' is the most important part.

Broom: A stiff lightweight yard or deck broom is the best. Look for one with a smallish head that can easily be manoeuvred in tight spaces.

Secateurs: There are many different designs and price points for secateurs including specialist ones for left-handers. The cleanness of the cut is all important so that any wound will heal over and stop the risk of disease creeping in. I prefer bypass secateurs, which cut like scissors, rather than anvil secateurs, which cut down against a flat surface.

Multi-change kits: Some manufacturers make kits with different 'heads' such as a hoe, rake and broom that can be attached to a single handle. I recommend them. They can be extremely space-saving and perfect for the 'occasional' gardener. If opting for this set-up, it's important to buy the best quality you can afford because the joints and fixing mechanism will get more wear than a conventional garden tool.

Extras

Trugs: Large plastic ones with handles are really useful for gathering debris or filling up with compost to distribute around the garden.

Pruners: Long-arm pruners will give you more power and reach than secateurs to get through thicker wood. Some have telescopic handles extending their reach further.

Pressure washer: Shady gardens and slippery surfaces are best pressure washed annually (without the need for chemicals). They can be hired but you may want to invest in one. Weigh up between the sizes: powerful, larger ones take up more space but save time.

Ladder: A lightweight aluminium ladder is always useful for pruning, training climbers and clearing gutters.

Hose: Retractable hoses can now easily be neatly fixed to the wall near the tap.

Hedge shears/trimmers: May not be required but can be used for jobs other than hedge trimming such as clipping back climbers, topiary or cutting back grasses and perennials.

Bulb planter: Great for planting lots of bulbs (get a long-handled one for bulbs in lawns) as they take out a core, the bulb can be popped in and soil replaced on top.

index

A

Abelia 63
Acer (maple) 114, 194
acid-loving plants 95, 112, 116, 120, 122, 194
Agapanthus 63, 201
agave 204–5
ageratum 95
aggregates 23, 24, 30–1
Ajuga 107
   A. reptans 137
Akebia quinata 147
allium 177
alpines 66, 87
Amelanchier lamarckii 110
Anemone
   A. blanda 177
   A. x hybrida 'Honorine Jobert' 48, 128
   Japanese 48, 128
Angelica archangelica 49
Angel's trumpet 163
annuals 95, 98, 146, 155–64
Antirrhinum (snapdragon) 95, 159
apple see Malus
Arbutus unedo 62, 115
arctotis 95
armeria 87
Asarina barclayana 163

Asarum 51, 167
   A. europaeum 142
ash, mountain 112
Asian greens 108
Astelia 170
aster 130
Astrantia major 'Claret' 129
atmosphere 13, 37–8
aubretia 87
aucuba 194
azalea 95, 194

B

bacopa 107
balconies 63–4
bamboo 42, 167
banana 167, 169, 204
basil 108
bean, dwarf French 95
beech 194
beetroot 108
Bergenia 107
   B. 'Overture' 143
Betula (birch) 63, 111
bicycle stores 55–6, 66
bin stores 56
biodiversity 66, 86
birch (Betula) 111
   silver 63
Bishop's hat see Epimedium
bistort, red 130

P

pruning 44, 68, 185–6
  root pruning 200–1
  top pruning 200–1
Prunus 114
  P. lusitanica 'myrtifolia' 126
Pulmonaria 'Blue Ensign' 141
pulsatilla 87
purple top 134

R

radish 108
railway sleepers 24, 101–2
raised beds 99–102
rakes 210
recycling stores 56, 188
rheum 75
rhododendron 95, 194
risers 24
rock gardens 177
romantic gardens 18–20
roof gardens 57–63
roofs, green 95–90
root pruning 200–1
rose 20, 42, 96, 119
  climbing 150–1
rosemary 63
rowan 112
Rudbeckia 157
runoff 25
Russian vine 40, 147

S

salad leaves 95, 108

Salvia (sage) 107, 108, 167
  S. guaranitica 'Amistad' 132
Sambucus nigra 42
  S. n. 'Black Beauty' 119
Sarcococca (sweet/Christmas box) 68
  S. confusa 46
  S. hookeriana var. *dignya* 'Purple Stem' 124
saxifrage 87
scabious 107
scale 14
scilla 177
screening 40–2, 62
secateurs 185, 201, 212
sedum 87, 107
  matting 90
sempervivum 87
shade 37–8, 40, 43–52
  dappled 44
  deep 44
  dry 44, 45–6, 48, 52, 136
  plants for 46–52, 107
shadowing 84
sheds 54, 86
shrubs 117–26
  and autumn colour 120–2
  climbing 67
  and containers 95, 96
  evergreen 42, 123–4, 126
  and front gardens 68
  and privacy 42
  and shade 44–7
  spring-flowering 118–19
  summer-flowering 118–19
  and topiary 126
  and winter scent 123–5